305708817$

D1576394

Multilevel Statistical Models

The development of modern statistical theory is reflected in the history of the late Sir Maurice Kendall's volumes, *The Advanced Theory of Statistics*. This landmark publication began life as a two volume work (Volume 1, 1943; Volume 2, 1946) and grew steadily, as a single authored work, until the late 1950s. At this stage, Alan Stuart became co-author and *The Advanced Theory* was rewritten in three volumes. When Keith Ord joined the authorial team in the early 1980s, plans were laid to expand still further to a four-volume work. However, it became evident that with the ongoing and often rapid pace of developments in statistics there were gaps in the coverage and that it was becoming increasingly difficult to provide timely updates to all volumes, so a new strategy was devised.

The Advanced Theory now forms a core of three volumes in an expanded series of related works, *The Kendall's Library of Statistics*. The other books in this series encompass the areas previously appearing in the old Volume 3 and to some extent in Volume 2 and the series structure allows the inclusion of new titles in other areas which have become established since the publication of the original *Advanced Theory*.

In the preface to the first edition of *The Advanced Theory*, Kendall declared that his aim was 'to develop a systematic treatment of [statistical] theory as it exists at the present time'. This continues to be the guiding principle of the series, which hopes to maintain its role as a standard and comprehensive reference on modern statistics.

KENDALL'S ADVANCED THEORY OF STATISTICS

Kendall's Advanced Theory of Statistics, Volume 1: Distribution Theory, 6th edn (1994), Stuart & Ord

Kendall's Advanced Theory of Statistics, Volume 2A: Classical Inference and the Linear Model (1999), Stuart, Ord & Arnold

Kendall's Advanced Theory of Statistics, Volume 2B: Bayesian Inference (1994), O'Hagan

KENDALL'S LIBRARY OF STATISTICS

Multivariate Analysis, Part 1: Distributions, Ordination and Inference (1994), Krzanowski & Marriott

Multivariate Analysis, Part 2: Classification, Covariance Structures and Repeated Measurements (1995), Krzanowski & Marriott

Multilevel Statistical Models, 3rd edn (2003), Goldstein

The Analysis of Proximity Data (1997), Everitt and Rabe-Hesketh

Robust Nonparametric Statistical Methods (1998), Hettmansperger & McKean

Statistical Regression with Measurement Error (1999), Cheng & Van Ness

Latent Variable Models and Factor Analysis (1999), Bartholomew & Knott

Statistical Inference for Diffusion Type Processes (1999), Prakasa Rao

Statistical Decision Theory (2000), French and Ríos Insua

For further details please see our website at www.kendallslibrary.com

MULTILEVEL STATISTICAL MODELS

Third Edition

Harvey Goldstein

Institute of Education, University of London, UK

A member of the Hodder Headline Group

LONDON

Distributed in the United States of America by
Oxford University Press Inc., New York

First published in Great Britain in 1987 as Multilevel
Models in Educational and Social Research

Second edition published in 1995 as Multilevel
Statistical Models by Edward Arnold
Reprinted in 1995, 1996(2), 1998, 1999, 2000

Third edition published in Great Britain in 2003 by
Arnold, a member of the Hodder Headline Group,
338 Euston Road, London NW1 3BH

http://www.arnoldpublishers.com

Distributed in the United States of America by
Oxford University Press Inc.,
198 Madison Avenue, New York, NY10016

British Library Cataloguing in Publication Data
A catalogue record for this book is available from the British Library

Library of Congress Cataloging-in-Publication Data
A catalog record for this book is available from the Library of Congress

ISBN 0 340 80655 9 (hb)

1 2 3 4 5 6 7 8 9 10

Production Controller: Martin Kerans
Cover Design: Terry Griffiths

Typeset in 10/11 pt Times by Charon Tec Pvt. Ltd, Chennai, India
Printed and bound in Great Britain by MPG Books Ltd, Bodmin, Cornwall

What do you think about this book? Or any other Arnold title?
Please send your comments to feedback.arnold@hodder.co.uk

Contents

Preface

In the mid 1980s a number of researchers began to see how to introduce systematic approaches to the statistical modelling and analysis of hierarchically structured data. The early work of Aitkin et al. (1981) on the teaching styles' data and Aitkin's subsequent work with Longford (1986) initiated a series of developments that by the early 1990s had resulted in a core set of established techniques, experience and software packages that could be applied routinely. These methods and further extensions of them are described in this book and are now applied widely in areas such as education, epidemiology, geography, child growth, household surveys and many others.

In addition to the first and second editions of the present text (Goldstein, 1987b, 1995), several expository volumes have now appeared (see section 1.14 in the introductory chapter). The present text aims to integrate existing methodological developments within a consistent terminology and notation, provide examples and explain a number of new developments, especially in the areas of discrete response data, multiple membership structures, factor analysis, random cross-classifications, errors of measurement and missing data. In almost all cases these developments are the subject of continuing research.

The main text seeks to avoid undue statistical complexity, with methodological derivations occurring in appendices. Examples and diagrams are used where possible to illustrate the application of the techniques and references are given to other work. The book is intended to be suitable for graduate-level courses and as a general reference.

Harvey Goldstein
September 2002

Acknowledgements

This book would not have been possible without the support and dedication of all those who have worked on or been closely associated with the Centre for Multilevel Modelling (formerly the Multilevel Models Project) at the Institute of Education. Those who have contributed substantially to the present volume include Jon Rasbash, Min Yang, William Browne, Fiona Steele, Ian Plewis, Michael Healy and Toby Lewis. They have invested time and effort in correcting mistakes and making many useful suggestions. In addition there are a large number of friends and colleagues, too numerous to mention, who have also pointed out errors, suggested improvements and generally provided encouragement. Needless to say, any remaining obscurities or mistakes are entirely the author's responsibility. Sincere thanks also are due to the Economic and Social Research Council for their almost continuous support with project funding since 1986.

Notation

Examples use two-level models

Definition	Symbol
Response variable vector	Y
Explanatory variable design matrix	X
Fixed part explanatory variable design matrix for a single unit	X_{ij} for a level 1 unit X_j for a level 2 unit
Total residuals at each level for a 2-level model	$u_j = \sum_{h=0}^{q_2} u_{hj} z_{hj}^{(2)}$ $e_{ij} = \sum_{h=0}^{q_1} e_{hij} z_{hij}^{(1)}$
Explanatory variable design matrix for level 2 and level 1 random coefficients	$Z^{(2)}, \ Z^{(1)}$
Predicted value from fixed part of model	$\hat{y}_{ij} = X_{ij}\hat{\beta} = (X\hat{\beta})_{ij}$
Raw or total residual for level 1 unit	$\tilde{y}_{ij} = y_{ij} - \hat{y}_{ij}$
Mean raw residual for level 2 unit	$\tilde{y}_j = \dfrac{1}{n_j} \sum_{i=1}^{n_j} \tilde{y}_{ij}$
Observations (y) independently and identically distributed with specified density function (g)	$y \overset{iid}{\sim} g$

Estimated residual or posterior residual estimate $\quad \hat{u}_j, \hat{e}_{ij}$

Covariance matrix of random coefficients at level i $\quad \Omega_i, \Omega = \{\Omega_i\}$

Parentheses denoting vector or matrix of elements $\quad \{\}$

Covariance matrix of response vector for
k-level model $\quad V_k$ or just V

Contribution to covariance matrix of response
vector from level i for k-level model $\quad V_{k(i)}$, or just $V_{(i)}$

Direct sum of matrices $A_1, \ldots, A_k$ $\quad \displaystyle\bigoplus_{i=1}^{k} A_i$

Kronecker product of matrices A_1, A_2 $\quad A_1 \otimes A_2$

vec operator on matrix A $\quad vec(A)$

Level 2 cross-classification model; classifications
indexed by j_1, j_2 $\quad \begin{aligned} y_{i(j_1 j_2)} &= X_{i(j_1 j_2)}\beta + u_{1j_1} \\ &+ u_{2j_2} + e_{i(j_1 j_2)} \end{aligned}$

Multiple membership model; particular case of
a level 1 unit belonging to just two level 2 units.
j_1, j_2 index units in the same classification $\quad \begin{aligned} y_{i(j_1 j_2)} &= X_{i(j_1 j_2)}\beta + w_{1ij_1}u_{j_1} \\ &+ w_{2ij_2}u_{j_2} + e_{i(j_1 j_2)} \\ w_{1ij_1} &+ w_{2ij_2} = 1 \end{aligned}$

A general classification notation and diagram

Response measurements and random effects have a single subscript that identifies the unit. Random effects, and optionally responses, also have a single superscript that identifies the 'classification'. Thus, a variance components 2-level model, using the example of students and schools, can be written in standard notation as

$$y_{ij} = (X\beta)_{ij} + u_j + e_{ij}$$
$$u_j \sim N(0, \sigma_u^2), \quad e_{ij} \sim N(0, \sigma_e^2)$$

and in the general classification notation as

$$y_i^{(1)} = (X\beta)_i + u_{school(i)}^{(2)} + u_{student(i)}^{(1)} \quad school(i) \in (1, \ldots, J)$$
$$student(i) \in (1, \ldots, N) \quad u_{school(i)}^{(2)} \sim N(0, \sigma_{u(2)}^2)$$
$$u_{student(i)}^{(1)} \sim N(0, \sigma_{u(1)}^2) \quad i = 1, \ldots, N$$

By default the lowest level classification is (1) and where no ambiguity arises this superscript may be omitted. The term *school(i)* refers to the school that the i-th student belongs to, school being the classification (set of units) with superscript (2). The term *student(i)* refers to student i which is classification (1). Classifications can be general, including nesting relationships, crossings and multiple memberships.

Accompanying this notation is a classification diagram which has the following elements.

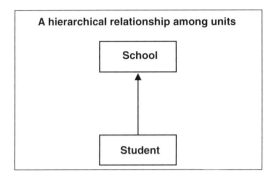

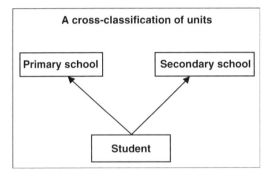

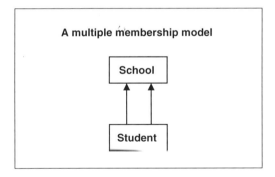

Glossary

Cluster	A grouping containing 'lower level' elements. For example in a sample survey the set of households in a neighbourhood.
Cross-classification	A structure where lower level units are grouped within the cells of a multiway classification of higher level units.
Design matrix	In the fixed part of the model, the matrix of values of the explanatory variables X. In the random part the matrix of explanatory variables Z.
Explanatory variable	Also known as an 'independent' variable. In the fixed part of the model usually denoted by x and in the random part by z.
Fixed part	That part of a model represented by $X\beta$, that is the average relationship.
Level	A component of a data hierarchy. Level 1 is the lowest level, for example students within schools or repeated measurement occasions within individual subjects.
Level n variation	The variation among level n unit measurements.
Multiple membership	A structure where a level unit may be nested within one or more higher level units.
Nesting	The clustering of units into a hierarchy
Random part	That part of a model represented by Zu, that is the contribution of the random variables, at each level.
Response variable	Also known as a 'dependent' variable. Denoted by y.
Unit	An entity defined at a level of a data hierarchy. For example an individual student will be a level 1 unit within a level 2 unit such as a school.

1

An Introduction to Multilevel Models

1.1 Hierarchically structured data

Many kinds of data, including observational data collected in the human and biological sciences, have a *hierarchical, nested,* or *clustered* structure. For example, animal and human studies of inheritance deal with a natural hierarchy where offspring are grouped within families. Offspring from the same parents tend to be more alike in their physical and mental characteristics than individuals chosen at random from the population at large. For instance, children from the same family may all tend to be small, perhaps because their parents are small or because of a common impoverished environment.

Many designed experiments also create data hierarchies, for example clinical trials carried out in several randomly chosen centres or groups of individuals. For now, we are concerned only with the *fact* of such hierarchies not their provenance. The principal applications we shall deal with are those from the social sciences, but the techniques are of course applicable more generally. In subsequent chapters, as we develop the theory and techniques with examples, we shall see how a proper recognition of these natural hierarchies allows us to seek more satisfactory answers to important questions.

We refer to a hierarchy as consisting of *units* grouped at different *levels*. Thus offspring may be the level 1 units in a 2-level structure where the level 2 units are the families: students may be the level 1 units clustered or nested within schools that are the level 2 units.

The existence of such data hierarchies is neither accidental nor ignorable. Individual people differ, as do individual animals, and this necessary differentiation is mirrored in all kinds of social activity where the latter is often a direct result of the former, for example when students with similar motivations or aptitudes are grouped in highly selective schools or colleges. In other cases, the groupings may arise for reasons less strongly associated with the characteristics of individuals, such as the allocation of young children to elementary schools, or the allocation of patients to different clinics. Once groupings are established, even if their establishment is effectively random, often they will tend to become differentiated, and this differentiation implies that the group and its members both influence and are influenced by the group membership. To ignore this relationship risks overlooking the importance of group effects, and may also render invalid many of the traditional statistical analysis techniques used for studying data relationships.

We shall be looking at this issue of statistical validity in the next chapter, but one simple example will show its importance. A well-known and influential study of primary (elementary) school children carried out in the 1970s (Bennett, 1976) claimed that children exposed to so called 'formal' styles of teaching reading exhibited more progress than those who were not. The data were analysed using traditional multiple regression techniques which recognized only the individual children as the units of analysis and ignored their groupings within teachers and into classes. The results were statistically significant. Subsequently, Aitkin et al. (1981) demonstrated that when the analysis accounted properly for the grouping of children into classes, the significant differences disappeared and the 'formally' taught children could not be shown to differ from the others.

This reanalysis is the first important example of a *multilevel* analysis of social science data. In essence what was occurring here was that the children within any one classroom, because they were taught together, tended to be similar in their performance. As a result they provided rather less information than would have been the case if the same number of students had been taught separately by different teachers. In other words, the basic unit for purposes of comparison should have been the teacher not the student. The function of the students can be seen as providing, for each teacher, an estimate of that teacher's effectiveness. Increasing the number of students per teacher would increase the precision of those estimates but not change the number of teachers being compared. Beyond a certain point, simply increasing the numbers of students in this way hardly improves things at all. However, increasing the number of teachers to be compared, with the same or somewhat smaller number of students per teacher, considerably improves the precision of the comparisons.

Researchers have long recognized this issue. In education, for example, there has been much debate (see Burstein et al., 1980) about the so-called 'unit of analysis' problem, that is the one just outlined. Before multilevel modelling became well developed as a research tool, the problems of ignoring hierarchical structures were reasonably well understood, but they were difficult to solve because powerful general purpose tools were unavailable. Special-purpose software, for example for the analysis of genetic data, has been available longer but this was restricted to 'variance components' models (see Chapter 2) and was not suitable for handling general linear models. Sample survey workers have recognized this issue in another form. When population surveys are carried out, the sample design typically mirrors the hierarchical population structure, in terms of geography and household membership. Elaborate procedures have been developed to take such structures into account when carrying out statistical analyses. We look at this in more detail in Chapter 9.

In the remainder of this chapter we shall look at the major areas explored in this book.

1.2   School effectiveness

Schooling systems present an obvious example of a hierarchical structure, with pupils clustered within schools, which themselves may be clustered within education authorities or boards. Educational researchers have been interested in comparing schools and other educational institutions, most often in terms of the achievements of their pupils. Such comparisons have several aims, including the aim of public accountability

(Goldstein, 1997) but, in research terms, interest usually is focused upon studying the factors that explain school differences.

Consider the common example where test or examination results at the end of a period of schooling are collected for each school in a randomly chosen sample of schools. The researcher wants to know whether a particular kind of subject streaming practice in some schools is associated with improved examination performance. She also has good measures of the pupils' achievements when they started the period of schooling so that she can control for this in the analysis. The traditional approach to the analysis of these data would be to carry out a regression analysis, using performance score as the response, to study the relationship with streaming practice, adjusting for the initial achievements. This is very similar to the initial teaching styles analysis described in the previous section, and suffers from the same lack of validity through failing to take account of the school level clustering of students.

An analysis that explicitly models the manner in which students are grouped within schools has several advantages. Firstly, it enables data analysts to obtain statistically efficient estimates of regression coefficients. Secondly, by using the clustering information it provides correct standard errors, confidence intervals and significance tests, and these generally will be more 'conservative' than the traditional ones that are obtained simply by ignoring the presence of clustering – just as Bennett's previously statistically significant results became non-significant on reanalysis. Thirdly, by allowing the use of covariates measured at any of the levels of a hierarchy, it enables the researcher to explore the extent to which differences in average examination results between schools are accountable for by factors such as organizational practice or possibly in terms of other characteristics of the students. It also makes it possible to study the extent to which schools differ for different kinds of students, for example to see whether the variation between schools is greater for initially high scoring students than for initially low scoring students (Goldstein et al., 1993) and whether some factors are better at accounting for or 'explaining' the variation for the former students than for the latter. Finally, there may be interest in the relative ranking of individual schools, using the performances of their students after adjusting for intake achievements. This can be done straightforwardly using a multilevel modelling approach and we shall see an example in Chapter 2.

To fix the basic notion of a level and a unit, consider Figures 1.1 and 1.2 based on hypothetical relationships. Figure 1.1 shows the exam score and intake achievement scores for five students in a school, together with a simple regression line fitted to the data points. The residual variation in the exam scores about this line is the *level 1 residual variation*, since it relates to level 1 units (students) within a sample level 2 unit (school). In Figure 1.2 the three lines are the simple regression lines for three schools, with the individual student data points removed. These vary in both their slopes and their intercepts (where they would cross the exam axis), and this variation is *level 2 variation*. It is an example of complex level 2 variation since both the intercept and slope parameters vary.

The other extreme to an analysis which ignores the hierarchical structure is one which treats each school completely separately by fitting a different regression model within each one. In some circumstances, for example where we have very few schools and moderately large numbers of students in each, this may be efficient. It may also be appropriate if we are interested in making inferences about just those schools. If, however, we regard these schools as a (random) sample from a population of schools and we wish to make inferences about the variation between schools in general, then a

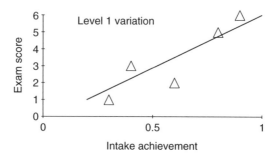

Figure 1.1   An illustration of level 1 variation.

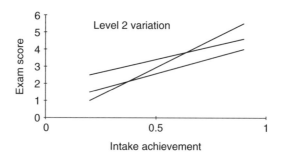

Figure 1.2   An illustration of level 2 variation.

full multilevel approach is called for. Likewise, if some of our schools have very few students, fitting a separate model for each of these will not yield reliable estimates: we can obtain more precision by regarding the schools as a sample from a population and using the information available from the whole sample data when making estimates for any one school. This approach is especially important in the case of repeated measures data where we typically have very few level 1 units per level 2 unit.

We introduce the basic procedures for fitting multilevel models to hierarchically structured data in Chapter 2 and discuss the design problem of choosing the numbers of units at each level in Chapter 3.

1.3   Sample survey methods

We have already mentioned sample survey data. The standard literature on surveys, reflected in survey practice, recognizes the importance of taking account of the clustering in complex sample designs. Thus, in a household survey, the first stage sampling unit will often be a well-defined geographical unit. From those that are randomly chosen, further stages of random selection are carried out until the final households are selected. Because of the geographical clustering exhibited by measures such as political attitudes, special procedures have been developed to produce valid statistical inferences, for example when comparing mean values or fitting regression models (Skinner et al., 1989).

While such procedures usually have been regarded as necessary they have not generally merited serious substantive interest. In other words, the population structure, insofar as it is mirrored in the sampling design, is seen as a 'nuisance factor'. By contrast, the multilevel modelling approach views the population structure as of potential interest in itself, so that a sample designed to reflect that structure is not merely a matter of saving costs as in traditional survey design, but can be used to collect and analyse data about the higher level units in the population.

Although the direct modelling of clustered data is statistically efficient, it will generally be important to incorporate weightings in the analysis which reflect the sample design or, for example, patterns of non-response, so that robust population estimates can be obtained and so that there will be some protection against serious model misspecification. A procedure for introducing external unit weights into a multilevel analysis is discussed in Chapter 3 and a discussion of analysing survey data is given in Chapter 9.

1.4 Repeated measures data

A different example of hierarchically structured data occurs when the same individuals or units are measured on more than one occasion. A common example occurs in studies of animal and human growth. Here the occasions are clustered within individuals that represent the level 2 units with measurement occasions as the level 1 units. Such structures are typically strong hierarchies because there is much more variation between individuals in general than between occasions within individuals. In the case of child height growth, for example, once we have adjusted for the overall trend with age, the variance between successive measurements on the same individual is generally no more than 5% of the variation in height between children.

There is a considerable literature on procedures for the analysis of such repeated measurement data (see for example Goldstein, 1979), which has more or less successfully confronted the statistical problems. It has done so, however, by requiring that the data conform to a particular, balanced, structure. Broadly these procedures require that the measurement occasions are the same for each individual. This may be possible to arrange, but often in practice individuals will be measured irregularly, some of them a great number of times and some perhaps only once. By considering such data as a general 2-level structure we can apply the standard set of multilevel modelling techniques that allow any pattern of measurements while providing statistically efficient parameter estimation. At the same time modelling a 2-level structure presents a simpler conceptual understanding of such data and leads to a number of interesting extensions that will be explored in Chapter 5.

One particularly important extension occurs in the study of growth where the aim is to fit growth curves to measurements over time. In a multilevel framework this involves, in the simplest case, each individual having their own straight line growth trajectory with the intercept and slope coefficients varying between individuals (level 2). When the level 1 measurements, considered as deviations from each individual's fitted growth curve, are not independent but have an autocorrelated or time series structure, neither the traditional procedures nor the basic multilevel ones are adequate. This situation may occur, for example, when measurements are made very close together in time so that a 'positive' deviation from the curve at one time implies also a positive deviation after the short interval before the next measurement. Chapter 5 will explore methods for handling such data.

1.5   Event history models

Modelling time spent in various states or situations is important in a number of areas. In industry the 'time to failure' of components is a key factor in quality control. In medicine the survival time is a fundamental measurement in studying certain diseases. In economics the duration of employment periods is of great interest. In education, researchers often study the time students spend on different tasks or activities.

In studying employment histories, any one individual will generally pass through several periods of employment or unemployment, while at the same time changing his or her characteristics, for example his or her level of qualifications. From a modelling point of view we need to model the length of time in each type of employment, relating this to both constant factors such as an individual's social origins or gender, and to changing or time-dependent factors such as qualifications and age. In this case the multilevel structure is analogous to that for repeated measures data, with periods taking the place of occasions. Furthermore, generally we would have a further, higher level of the hierarchy, since individuals, which are the level 2 units, are themselves typically clustered into workplaces, which now constitute level 3 units.[1] In fact, the structure may be even more complicated if these workplaces change from period to period, and if we wish to include this level in our model we need to consider cross-classifications of the units (see below). Particular problems arise when studying event duration data that are encountered when some information is 'censored' in the sense that instead of being able to observe the actual duration we only know that it is longer than some particular value, or in some cases less than a particular value. Chapter 10 will discuss ways of dealing with multilevel event history models in detail.

1.6   Discrete response data

Until now we have assumed implicitly that our response or dependent variable is continuously distributed, for example an exam score or anthropometric measure such as height. Many kinds of statistical modelling, however, deal with categorized responses, in the simplest case with proportions. Thus, we might be interested in a mortality rate, or an exam pass rate and how these vary from area to area or school to school.

In studying mortality rates in a population, it is often of great concern to try to understand the factors associated with variations from area to area or community to community. This produces a basic 2-level structure with individuals at level 1 and communities at level 2. A typical study might record deaths over a given time period together with the characteristics of the individuals concerned, along with a control group and level 2 characteristics of the communities, such as their sizes or social compositions. One analysis of interest would be to see whether any of these explanatory variables could explain between-community variation. Another interest might be in studying whether mortality rate differences, say between men and women, varied from community to community.

---

[1] Formally, we can regard unemployment for this purpose as a particular workplace.

Such models, part of the class known as generalized linear models, have been available for some time for single-level data (McCullagh and Nelder, 1989), with associated software. In Chapter 4 we show how to fit multilevel models with several categorical responses and even models with mixtures of categorical and continuous responses.

1.7 Multivariate models

An interesting special case of a 2-level model is the multivariate linear (or generalized linear) model. Suppose we have taken several measurements on an individual, for example their systolic and diastolic blood pressure and their heart rate. If we wish to analyse these together as response variables we can do so by setting up a multivariate, in this case three-variate, model with explanatory variables such as age, gender, social background, smoking exposure, etc. We can think of this as a 2-level model by considering each individual as a level 2 unit, with the three measurements constituting the level 1 units, rather as occasions did for the repeated measures model. Chapter 6 will show how this formal device for specifying a multivariate model yields considerable benefits. For example, by considering further higher levels, in this case say clinics, we have a simple way of specifying a multivariate multilevel model. Also, if some individuals do not have all the measurements, for example if they are randomly missing a blood pressure measurement, then this is automatically taken account of in the analysis, without the need for special procedures for handling missing data.

A particularly important application occurs where measurements are missing by design rather than at random. In certain kinds of surveys, known as rotation designs, and in certain kinds of educational assessments, known as matrix sample designs, each individual unit has only a subset of measurements made on it. For example, in large-scale testing programmes, the full range of tests may be too extensive for any one student, so that each student responds to only one combination. Such designs are viewed usefully as having a multivariate response, with the full set of tests constituting the complete multivariate response vector, and every student having some tests missing. Such designs can become rather complex, especially since the students themselves are clustered into schools. By viewing the data as a single hierarchy in which the multivariate responses are level 1, we obtain an efficient and readily interpretable analysis.

The multivariate multilevel model can also be used as the basis for one approach to dealing with missing data in multilevel models and this is developed in Chapter 14.

1.8 Nonlinear models

Some kinds of data are better represented in terms of nonlinear rather than linear models. For example, the modelling of discrete response data is considered formally as a case of modelling nonlinear data. Many kinds of growth data are conveniently modelled in this way, especially during periods of rapid and complex growth such as early infancy and at the approach to adulthood when growth approaches an upper asymptote (Goldstein, 1979). Other examples arise when the response variable has inherent constraints. For example, biochemical activity patterns in patients may exhibit asymptotic behaviour, or cyclical patterns, both of which are difficult to model using purely linear models. Chapter 8 will introduce such models and show how to extend the linear multilevel model to this case. It will also consider cases where variances and covariances can be modelled as nonlinear functions of explanatory variables.

1.9   Measurement errors

Most measurements made in the human sciences contain some error component. This may be due to observer error as when measuring the weight of an animal, or an inherent result of being able to measure only a small sample of behaviour as in educational testing. It is well known that when variables in statistical models contain relatively large components of such error, the resulting statistical inferences can be very misleading unless careful adjustments are made (Fuller, 1987). In the case of simple regression, when the explanatory or independent variable is measured with error, the usual estimate of the regression line slope is an underestimate compared to that which would result if the measurement were available without error. This is particularly important, for example, in studies of school effectiveness where the fitting of intake achievement scores is important but where such scores often have large components of measurement error.

An important case when the latter arise is where the level 2 variable is a 'compositional' variable. That is, it is a measurement aggregated from the characteristics of the level 1 units within the level 2 units. Thus for example the mean intake achievement and the standard deviation of the intake achievements of all the pupils in a school are compositional variables that may, and indeed sometimes do, affect the final achievements of each individual student. Likewise in a household survey, we may consider that a measure of the average social status or the percentages of households in each social group, using all the households in the immediate community, are important explanatory variables to fit in a model. The problem arises when it is possible to collect data on only some of the level 1 units, this often being the case with household sample surveys. What we then have is an estimate of a compositional variable that is measured with error, in the case of household surveys typically with a very large error. In many educational studies this also occurs where only a small proportion of students within a class or school are sampled.

Chapter 13 discusses the problems of dealing with measurement errors in multilevel models.

1.10   Random cross-classifications and multiple membership structures

We have already alluded to examples where units are cross-classified as well as clustered. In geographical research, the definition of an individual's geographical area is contingent upon the context being considered. Thus, the relevant location unit for purposes of leisure may not be the same as that surrounding the environment of work or schooling. We can conceive formally of individuals belonging simultaneously to both types of unit, each of which may have an influence on a person's life.

In most schooling systems, students move from elementary to secondary or high school. We might expect that both the elementary and secondary schools attended will influence a student's achievements or attitudes measured at the end of secondary school. Thus the level 2 units are of two types, elementary school and secondary school, where each 'cell' of their cross-classification contains some, or possibly no students. In this example, a third way of classification could be the area or neighbourhood where the student lives. Chapter 11 explores such cross-classified structures.

An interesting situation occurs where for a single level 2 classification, level 1 units may belong to more than one level 2 unit. An example from sociology concerns children's and adults' friendship patterns where an individual may belong to several groups simultaneously. The characteristics of the members of each group will influence such an individual, in relation to the individual's exposure to the group. In a longitudinal study of schooling, many students will change schools during the course of the study. The contribution to the response from schools will therefore reflect, for these students, the 'effect' of every school they have attended. With a suitable set of weights to reflect the time spent in each school this can be taken into account in the analysis. Such 'multiple membership models' are discussed in Chapter 12.

To handle the complexity of multiple membership and cross-classified structures, as well as mixtures of these, a special notation and set of diagrams will be introduced that allows a complete specification of such models.

1.11   Factor analysis and structural equation models

In many areas of the social sciences, where measurements are difficult to define precisely, an investigator might suppose that there is some underlying construct which cannot be measured directly but nevertheless can be assessed indirectly by measuring a number of relevant indicators. Structural equation modelling, and in particular the special case of factor analysis, was developed for this purpose, typically dealing with individuals' behaviour, attitudes or mental performance. Where individuals are grouped within hierarchies, for all the same reasons discussed above, it is important to carry out such analyses in a multilevel framework. For example, we may be interested in underlying individual attitudes based upon a number of indicators. Data on such indicators may be available over time and we can postulate a model whereby the underlying attitude varies from individual to individual (level 2) and also varies randomly over time within individuals (level 1). The model can then be further elaborated by studying whether there is any systematic change over time and whether this varies across individuals. Chapter 7 discusses such models.

1.12   Levels of aggregation and ecological fallacies

When studying relationships among variables, there has often been controversy about the appropriate 'unit of analysis'. We have alluded to this already in the context of ignoring hierarchical data clustering and, as we have seen, the issue is resolved by explicit hierarchical modelling.

One of the best known early illustrations of what is often known as the ecological or aggregation fallacy was the study by Robinson (1950) of the relationship between literacy and ethnic background in the United States. When the mean literacy rates and mean proportions of black Americans for each of nine census divisions are correlated the resulting value is 0.95, whereas the individual-level correlation ignoring the grouping is 0.20. Robinson was concerned to point out that aggregate-level relationships could not be used as estimates for the corresponding individual-level relationships and this point is now well understood. In Chapter 3 we shall discuss some of the statistical consequences of modelling only at the aggregate level.

Sometimes the aggregate level is the principal level of interest, but nevertheless a multilevel perspective is useful. Consider the example (Derbyshire, 1987) of predicting the proportion of children socially 'at risk' in each local administrative area for the purpose of allocating central government expenditure on social services. Survey data are available for individual children with information on risk status so that a prediction can be made using area-based variables as well as child- and household-based variables. The probability (π) of a child being 'at risk' was estimated by the following (single-level) equation:

$$\text{logit}(\pi) = -6.3 + 5.9x_1 + 2.2x_2 + 1.5x_3$$

where x_1 is the proportion of children in the area in households with a lone parent, x_2 is the proportion of households in each area which have a density of more than 1.5 persons per room and x_3 is the proportion of households whose 'head' was born in the British 'New Commonwealth' or Pakistan. All these explanatory variables are measured at the aggregate area level and the response is the proportion of children at risk in each area. Although we can regard this analysis as taking place entirely at the area level (with suitable weighting for the number of children in each area), there are advantages in thinking of it as a 2-level model with each child being a level 1 unit and the response variable being the binary response of whether or not the child is at risk.

Firstly, this allows us to incorporate possibly important variables that are measured at the child level, for example whether or not each child's household is overcrowded. Including such level 1 variables may greatly improve the predictive power of the model. With the results of such a model we can then form a prediction for each area by aggregating over the known numbers of children living in overcrowded households.

Secondly, the possibility of modelling the characteristics of children or their households allows the possibility of an allocation formula that can take account of costs and benefits related to the actual composition of each area in terms of these child characteristics.

1.13   Causality

In the natural sciences, experimentation has a dominant position when making causal inferences. This is both because the units of interest can be manipulated experimentally, typically using random allocation, and because there is a widespread acceptance that the results of experiments are generalizable over space and time. The models described in this book can be applied to experimental or non-experimental data, but the final causal inferences will differ. Nevertheless, most of the examples used are from non-experimental studies in the human sciences and a few words on causal inferences from such data may be useful.

If we wish to answer questions about a possible causal relationship between, say, class size and educational achievement, an experimental study would need to assign different numbers of level 1 units (students) randomly to level 2 units (classes or teachers) and study the results over a time period of several years. This would be time consuming and could create ethical problems. In addition to such practical problems, any single study would be limited in time and place, and require extensive replication before results confidently could be generalized. The specific context of any study is important, for example the state of the educational system and the resources available at the time of the study. The difficulty from an experimental viewpoint is that it is practically impossible to allocate randomly with respect to all such possible confounding factors.

A further limitation of randomized controlled trails (RCTs) is that they cannot necessarily deal with situations where the *composition* of a higher level unit interacts with the treatment of interest, to affect the responses of lower level units. Thus, in schooling studies the size of class may affect the progress of students only when the proportion of 'low-achieving' students is above a certain threshold. Randomization will tend to eliminate classes with extreme proportions so that such effects may not be discovered. Goldstein (1998) looks at this issue in detail

None of this is to say that randomized experiments should never be undertaken, rather that on their own they may have limited potential for making general statements about causality. Whether an experiment fails or succeeds in demonstrating a relationship, there will almost always be further explanations for the findings which require study. Even if an experiment appears to eliminate a possible relationship, for example demonstrating a negligible relationship between class size and attainment, it may be legitimate to query whether a relationship nevertheless exists for specific subgroups of the population.

In the pursuit of causal explanations we require some guiding underlying principles or theories. It is these which will tell us what kinds of things to measure and how to be critical of findings. For example, in studies of the relationship between perinatal mortality and maternal smoking in pregnancy (Goldstein, 1976), we can attempt to adjust for confounding factors, such as poverty, which may be responsible for influencing both smoking habits and mortality. We can also study how the relationship varies across groups and seek measures which explain such variation. We might also, in some circumstances, be able to carry out randomized experiments, assigning for example intensive health education to a randomly selected 'treatment' group and comparing mortality rates with a 'control' group.

A multilevel approach could be useful here in two different ways. Firstly, pregnant women will be grouped hierarchically, geographically and by medical institution and the between-area and between-institution variation may affect mortality and the relationship between mortality and smoking. Secondly, we will be able often to obtain serial measurements of smoking, so allowing the kind of repeated measures 2-level modelling discussed earlier. This will allow us to study how changes in smoking are related to mortality, and permit a more detailed exploration of possible causal mechanisms.

Multilevel models can often be used to identify units with extreme values. For example, in school effectiveness studies an exploration of school-level residual estimates (see Chapter 2) may identify those which are highly atypical, having adjusted for 'contextual' variables such as the intake characteristics of their students. These can then be selected for further scrutiny, for example by means of intensive case studies, so forming a link between the quantitatively based multilevel analysis and a more qualitatively based investigation which would seek to identify detailed causal processes.

A discussion of some necessary conditions for causal inference in observational studies can be found, for example, in Holland (1986) and Cochran (1983).

Finally, many of the concerns addressed by multilevel models are to do with prediction rather than causation. Thus, for example, in Chapter 5 we use a 2-level model of children's growth for the purpose of predicting adult height. In studies of school effectiveness we may be interested in understanding the causes of school differences, but we may be concerned also with predicting which school is likely to produce the best (on average) examination result for a student with given initial characteristics and achievements.

1.14   Other references

While the present volume aims to provide a comprehensive coverage of the topic of multilevel models, there are now many other texts which deal with specialized areas. Many of these are referenced in the appropriate chapters, but there are also several recent books which provide a good general introduction and/or very detailed worked examples. Among these are, Kreft and De Leeuw (1998), Snijders and Bosker (1999), Little et al. (2000), Heck and Thomas (2000), McCulloch and Searle (2001), Hox (2002) and Bryk and Raudenbush (2002). There are also edited collections of articles starting to appear on particular application areas; for example, Leyland and Goldstein (2001) bring together a collection of papers on the multilevel modelling of health statistics.

1.15   A caveat

The purpose of this book is to bring together techniques for the analysis of highly structured, multilevel data. The application of such techniques has already begun to yield new and important insights in a number of areas as the examples in the following chapters illustrate. As software becomes more widely available, the application of these techniques should become relatively straightforward, even routine.

All this is welcome, yet despite their usefulness, models for multilevel analysis cannot be a universal panacea. In some circumstances, where there is little structural complexity, they may be hardly necessary, and traditional single-level models may suffice, both for analysis and presentation. On the other hand multilevel analyses can bring extra precision to attempts to understand causality, for example by making efficient use of student achievement data in attempts to understand differences between schools. They are not, however, substitutes for well-grounded substantive theories, nor do they replace the need for careful thought about the purpose of any statistical modelling. Furthermore, by introducing more complexity they can extend but not necessarily simplify interpretations.

Multilevel models are tools to be used with care and understanding.

2

The Basic Two-Level Model

2.1 Introduction

In this chapter we introduce the 2-level model together with the basic notation which we shall use and develop throughout the book. We look at alternative ways of setting up and motivating the model and introduce procedures for estimating parameters, forming and testing functions of the parameters and constructing confidence intervals. We introduce alternative methods of estimation which will be used and elaborated upon in subsequent chapters.

To make matters concrete, consider the following data. It is a dataset we shall use again and it consists of 728 pupils in 48 primary (elementary) schools in inner London, part of the 'Junior School Project' (JSP). We consider two measurement occasions: the first when the pupils were in their fourth year of schooling, that is the year they attained their eighth birthday, and three years later in their final year of primary school. Our data are in fact a subsample from a more extensive dataset which is described in detail in Mortimore et al. (1988). We use the scores from mathematics tests administered on these two occasions together with information collected on the social background of the pupils and their gender. In this chapter the data are used primarily to illustrate the development of basic 2-level modelling. In Chapter 3 we shall be studying more elaborate models which will enable us to handle these data more efficiently.

Figure 2.1 is a scatterplot of the 11-year-old mathematics test score by the 8-year-old test score. In this plot no distinction is made between the schools to which the pupils belong. Notice that there is a general trend, with increasing 8-year scores associated with increasing 11-year scores. Notice also the narrowing of the between-pupil variation in the 11-year score with increasing 8-year score; an issue to which we shall return.

In Figure 2.2 the scores for two particularly different schools have been selected, represented by different symbols. Two things are apparent immediately. The school represented by the circles shows a steeper 'slope' than the school represented by the filled triangles and for most 8-year scores, the 11-year scores tend to be lower. Both these features are now addressed by formally modelling these relationships.

Consider first a simple model for one school, relating 11-year score to 8-year score. We write

$$y_i = \alpha + \beta x_i + e_i \tag{2.1}$$

where standard interpretations can be given to the intercept (α), slope (β) and residual (e_i). We would also typically assume that the residuals follow a Normal distribution

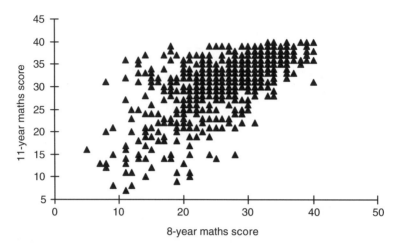

Figure 2.1   Scatterplot of 11-year by 8-year mathematics test scores. Some points represent more than one pupil.

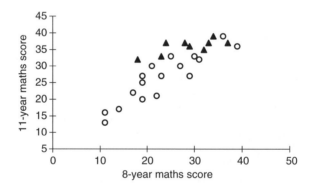

Figure 2.2   Scatterplot of 11-year by 8-year mathematics test scores for two schools.

with a zero mean and common variance, that is, $e_i \sim N(0, \sigma_e^2)$, but for now we shall concentrate on the variance properties of the residuals. We follow the normal convention of using Greek letters for the regression coefficients and place a circumflex over any coefficient (parameter) which is a sample estimate. This is the formal model for Figure 1.1 in the previous chapter and describes a single-level relationship. To describe simultaneously the relationships for several schools we write, for school j,

$$y_{ij} = \alpha_j + \beta_j x_{ij} + e_{ij} \tag{2.2}$$

This is now the formal model for Figure 1.2 where j refers to the level 2 unit (school) and i to the level 1 unit (pupil).

As it stands, (2.2) is still essentially a single-level model, albeit describing a separate relationship for each of the m schools. In some situations, for example where there are few schools and interest centres on just those schools in the sample, we may analyse (2.2) by fitting all the $2m + 1$ parameters, namely

$$(\alpha_j, \beta_j) \quad j = 1, \ldots, m \quad \sigma_e^2$$

assuming a common 'within-school' residual variance and separate lines for each school.

If we wish to focus not just on these schools, but on a wider 'population' of schools then we need to regard the chosen schools as giving us information about the characteristics of all the schools in the population. Just as we choose random samples of individuals to provide estimates of population means etc., so a randomly chosen sample of schools can provide information about the characteristics of the population of schools. In particular, such a sample can provide estimates of the variation and covariation between schools in the slope and intercept parameters and will allow us to compare schools with different characteristics.

An important class of situations arises when we wish primarily to have information about each individual school in a sample, but where we have a large number of schools so that (2.2) would involve estimating a very large number of parameters. Furthermore, some schools may have rather small numbers of students and application of (2.2) would result in imprecise and possibly widely fluctuating estimates. In such cases, if we regard the schools as members of a population and then use our population estimates of the mean and between-school variation, we can utilize this information to obtain more precise estimates for each individual school. This will be discussed later when we deal with 'residuals'.

2.2   The 2-level model

We now develop a general notation which will be used throughout this and later chapters, elaborated where necessary. We then discuss the estimation of model parameters and residuals and this is followed by illustrative examples.

To make (2.2) into a genuine 2-level model we let α_j and β_j become random variables. For consistency of notation replace α_j by β_{0j} and β_j by β_{1j} and assume that

$$\beta_{0j} = \beta_0 + u_{0j}, \quad \beta_{1j} = \beta_1 + u_{1j}$$

where u_{0j}, u_{1j} are random variables with parameters

$$E(u_{0j}) = E(u_{1j}) = 0$$

$$\text{var}(u_{0j}) = \sigma_{u0}^2, \quad \text{var}(u_{1j}) = \sigma_{u1}^2, \quad \text{cov}(u_{0j}, u_{1j}) = \sigma_{u01} \tag{2.3}$$

We can now write (2.2) in the form

$$y_{ij} = \beta_0 + \beta_1 x_{ij} + (u_{0j} + u_{1j} x_{ij} + e_{0ij})$$

$$\text{var}(e_{0ij}) = \sigma_{e0}^2 \tag{2.4}$$

And we shall require the extra suffix in the level 1 residual term for the models introduced in Chapter 3. We have expressed the response variable y_{ij} as the sum of a fixed part and a random part within the brackets and we shall also generally write the fixed part of (2.4) in the matrix form

$$E(Y) = X\beta \quad \text{with } Y = \{y_{ij}\}$$

$$E(y_{ij}) = X_{ij}\beta = (X\beta)_{ij}, \quad X = \{X_{ij}\}$$

where {} denotes a matrix, X is the design matrix for the explanatory variables and X_{ij} is the ij-th row of X. For model (2.4) we have $X = \{1 \ x_{ij}\}$. Note the alternative representation for the i-th row of the fixed part of the model.

The random variables are referred to as 'residuals' and in the case of a single-level model the level 1 residual e_{0ij} becomes the usual linear model residual term. To make the model equation symmetrical so that each coefficient has an associated explanatory variable, we can define a further explanatory variable for the intercept β_0 and its associated residual, u_{0j}, namely x_{0ij}, a constant which takes the value 1.0. For simplicity this variable sometimes may be omitted.

The feature of (2.4) which distinguishes it from standard linear models of the regression or analysis of variance type is the presence of more than one residual term and this implies that special procedures are required to obtain satisfactory parameter estimates. Note that it is the structure of the random part of the model which is the key factor. In the fixed part the variables can be measured at any level, for example in the JSP data we can measure characteristics of schools or teachers. We can also include so-called 'compositional' variables such as the average 8-year mathematics test score for all pupils in each school. The presence of such variables does not alter the estimation procedure, although results will require careful interpretation.

2.3   Parameter estimation

2.3.1   The variance components model

Equation (2.4) requires the estimation of two fixed coefficients, β_0 and β_1, and four other parameters, $\sigma_{u0}^2, \sigma_{u1}^2, \sigma_{u01}$ and σ_{e0}^2. We refer to such variances and covariances as *random parameters*. We start, however, by considering the simplest 2-level model which includes only the random parameters $\sigma_{u0}^2, \sigma_{e0}^2$. It is termed a variance components model because the variance of the response, about the fixed component, the *fixed predictor*, is

$$\text{var}(y_{ij}|\beta_0, \beta_1, x_{ij}) = \text{var}(u_0 + e_{0ij}) = \sigma_{u0}^2 + \sigma_{e0}^2$$

that is, the sum of a level 1 and a level 2 variance. For the JSP data this model implies that the total variance for each student is constant and that the covariance between two students (denoted by i_1, i_2) in the same school is given by

$$\text{cov}(u_{0j} + e_{0i_1 j}, u_{0j} + e_{i_2 j}) = \text{cov}(u_{0j}, u_{0j}) = \sigma_{u0}^2 \qquad (2.5)$$

since the level 1 residuals are assumed to be independent. The correlation between two such students is therefore

$$\rho = \frac{\sigma_{u0}^2}{(\sigma_{u0}^2 + \sigma_{e0}^2)}$$

$$\begin{pmatrix} \sigma_{u0}^2 + \sigma_{e0}^2 & \sigma_{u0}^2 & \sigma_{u0}^2 \\ \sigma_{u0}^2 & \sigma_{u0}^2 + \sigma_{e0}^2 & \sigma_{u0}^2 \\ \sigma_{u0}^2 & \sigma_{u0}^2 & \sigma_{u0}^2 + \sigma_{e0}^2 \end{pmatrix}$$

Figure 2.3 Covariance matrix of three students in a single school for a variance components model.

$$\begin{pmatrix} A & 0 \\ 0 & B \end{pmatrix}$$

where

$$A = \begin{pmatrix} \sigma_{u0}^2 + \sigma_{e0}^2 & \sigma_{u0}^2 & \sigma_{u0}^2 \\ \sigma_{u0}^2 & \sigma_{u0}^2 + \sigma_{e0}^2 & \sigma_{u0}^2 \\ \sigma_{u0}^2 & \sigma_{u0}^2 & \sigma_{u0}^2 + \sigma_{e0}^2 \end{pmatrix}$$

$$B = \begin{pmatrix} \sigma_{u0}^2 + \sigma_{e0}^2 & \sigma_{u0}^2 \\ \sigma_{u0}^2 & \sigma_{u0}^2 + \sigma_{e0}^2 \end{pmatrix}$$

Figure 2.4 The block-diagonal covariance matrix for the response vector Y for a 2-level variance components model with two level 2 units.

which is referred to as the 'intra-level-2-unit correlation'; in this case the intra-school correlation.[1] For the variance components model this also measures the proportion of the total variance which is between schools. In a model with three levels, say with schools, classrooms and students, we will have two such correlations and variance proportions; the intra-school correlation which is also the proportion of variance that is between schools and the intra-classroom correlation which is also that between classrooms. In more complex models with random coefficients the intra-unit correlation is not equivalent to the proportion of variance at the higher level and we shall look at this further in Chapter 3. To avoid confusion we shall use the term *variance partition coefficient* (VPC) to describe the proportion of variance at level 2 or at higher levels.

The existence of a non-zero intra-unit correlation, resulting from the presence of more than one residual term in the model, means that traditional estimation procedures such as 'ordinary least squares' (OLS) which are used for example in multiple regression, are inapplicable. A later section illustrates how the application of OLS techniques can lead to incorrect inferences. We now look in more detail at the structure of a 2-level data set, focusing on the covariance structure typified by Figure 2.3.

The matrix in Figure 2.3 is the (3×3) covariance matrix for the scores of three students in a single school, derived from the above expressions. For two schools, one with three students and one with two, the overall covariance matrix is shown in

---

[1] In the sample survey literature and elsewhere such as in genetics, the term 'intra-class correlation' is used, but this clearly is confusing in the present context.

$$V_2 = \begin{bmatrix} \sigma_{u0}^2 J_{(3)} + \sigma_{e0}^2 I_{(3)} & 0 \\ 0 & \sigma_{u0}^2 J_{(2)} + \sigma_{e0}^2 I_{(2)} \end{bmatrix}$$

Figure 2.5   Block-diagonal covariance matrix using general notation.

Figure 2.4 . This 'block-diagonal' structure reflects the fact that the covariance between students in different schools is zero, and clearly extends to any number of level 2 units.

A more compact way of presenting this matrix, which we shall use again, is given in Figure 2.5 where $I_{(n)}$ is the (n × n) identity matrix and $J_{(n)}$ is the (n × n) matrix of ones. The subscript 2 for V indicates a 2-level model. In single-level OLS models σ_{u0}^2 is zero and this covariance matrix then reduces to the standard form $\sigma^2 I$ where σ^2 is the (single) residual variance.

2.3.2   The general 2-level model with random coefficients

We can extend (2.4) in the standard way to include further fixed explanatory variables

$$y_{ij} = \beta_0 + \beta_1 x_{1ij} + \sum_{h=2}^{p} \beta_h x_{hij} + (u_{0j} + u_{1j} x_{1ij} + e_{0ij})$$

and more compactly as

$$y_{ij} = X_{ij}\beta + \sum_{h=0}^{1} u_{hj} z_{hij} + e_{0ij} z_{0ij} \tag{2.6}$$

where we use new explanatory variables for the random part of the model and write these more generally as

$$Z = \{Z_0\, Z_1\}$$
$$Z_0 = \{1\} \text{ i.e. a vector of 1's}$$
$$Z_1 = \{x_{1ij}\}$$

The explanatory variables for the random part of the model are often a subset of those in the fixed part, as here, but this is not necessary and later we shall encounter cases where this is not so. Also, any of the explanatory variables may be measured at any of the levels; for example we may have student characteristics at level 1 or school characteristics at level 2. Examples of both are used in the data analysis in a later section.

This model, with the coefficient of X_1 random at level 2, gives rise to the following typical block structure, for a level 2 block with two level 1 units. The matrix Ω_2 is the covariance matrix of the random intercept and slope at level 2. Note that we need to distinguish carefully between the covariance matrix of the responses given in Figure 2.6 and the covariance matrix of the random coefficients. We also refer to the intercept as a random coefficient. The matrix Ω_1 is the covariance matrix for the set of level 1 random coefficients; in this case there is just a single variance term at level 1. We will also write $\Omega = \{\Omega_i\}$ for the set of these covariance matrices.

We also see here the general pattern for constructing the response covariance matrix which generalizes both to higher order models and, as we shall see in Chapter 3, to complex variation structures at level 1.

$$\begin{pmatrix} A & B \\ B & C \end{pmatrix}$$

where

$$A = \left(\sigma_{u0}^2 + 2\sigma_{u01}x_{1j} + \sigma_{u1}^2 x_{1j}^2 + \sigma_{e0}^2\right)$$
$$B = \left(\sigma_{u0}^2 + \sigma_{u01}(x_{1j} + x_{2j}) + \sigma_{u1}^2 x_{1j}x_{2j}\right)$$
$$C = \left(\sigma_{u0}^2 + 2\sigma_{u01}x_{2j} + \sigma_{u1}^2 x_{2j}^2 + \sigma_{e0}^2\right)$$

giving

$$\begin{pmatrix} A & B \\ B & C \end{pmatrix} = X_j \Omega_2 X_j^T + \begin{pmatrix} \Omega_1 & 0 \\ 0 & \Omega_1 \end{pmatrix}$$

$$X_j = \begin{pmatrix} 1 & x_{1j} \\ 1 & x_{2j} \end{pmatrix}, \quad \Omega_2 = \begin{pmatrix} \sigma_{u0}^2 & \sigma_{u01} \\ \sigma_{u01} & \sigma_{u1}^2 \end{pmatrix}, \quad \Omega_1 = \sigma_{e0}^2$$

Figure 2.6 Response covariance matrix for a level 2 unit with two level 1 units for a 2-level model with a random intercept and random regression coefficient at level 2.

We now present a basic maximum likelihood (ML) procedure for obtaining estimates for our models before continuing to explore the dataset. Later in the chapter we will look at other estimation procedures.

2.4 Maximum likelihood estimation using Iterative Generalized Least Squares (IGLS)

The IGLS algorithm forms the basis for many of the developments in later chapters and we now summarize the main features. Appendix 2.1 sets out the details.

Consider the simple 2-level variance components model

$$y_{ij} = \beta_0 + \beta_1 x_{ij} + u_{0j} + e_{0ij}, \quad e_{0ij} \sim N(0, \sigma_{e0}^2) \tag{2.7}$$

Suppose that we knew the values of the variances, and so could construct immediately the block-diagonal matrix V_2, which we will refer to simply as V. We can then apply immediately the usual Generalized Least Squares (GLS) estimation procedure to obtain the estimator for the fixed coefficients

$$\hat{\beta} = (X^T V^{-1} X)^{-1} X^T V^{-1} Y \tag{2.8}$$

where in this case

$$X = \begin{pmatrix} 1 & x_{11} \\ 1 & x_{21} \\ \vdots & \vdots \\ 1 & x_{n_m m} \end{pmatrix} \quad Y = \begin{pmatrix} y_{11} \\ y_{21} \\ \vdots \\ y_{n_m m} \end{pmatrix} \tag{2.9}$$

with m level 2 units and n_j level 1 units in the j-th level 2 unit. Since we assume that the residuals have Normal distributions (2.8) also yields maximum likelihood estimates.

Our estimation procedure is iterative. We would usually start from 'reasonable' estimates of the fixed parameters. Typically these will be those from an initial OLS fit (that is assuming $\sigma_{u0}^2 = 0$), to give the OLS estimates of the fixed coefficients $\hat{\beta}^{OLS}$. From these we form the 'raw' residuals

$$\tilde{y}_{ij} = y_{ij} - \hat{\beta}_0^{OLS} - \hat{\beta}_1^{OLS} x_{ij} \tag{2.10}$$

The vector of raw residuals is written

$$\tilde{Y} = \{\tilde{y}_{ij}\}$$

If we form the cross-product matrix $\tilde{Y}\tilde{Y}^T$ we see that the expected value of this is simply V. We can rearrange this cross-product matrix as a vector by stacking the columns one on top of the other, which is written as $vec(\tilde{Y}\tilde{Y}^T)$ and similarly we can construct the vector $vec(V)$. For the structure given in Figure 2.4 these both have $3^2 + 2^2 = 13$ elements. The relationship between these vectors can be expressed as the following linear model

$$\begin{pmatrix} \tilde{y}_{11}^2 \\ \tilde{y}_{21}\tilde{y}_{11} \\ \vdots \\ \tilde{y}_{22}^2 \end{pmatrix} = \begin{pmatrix} \sigma_{u0}^2 + \sigma_{e0}^2 \\ \sigma_{u0}^2 \\ \vdots \\ \sigma_{u0}^2 + \sigma_{e0}^2 \end{pmatrix} + R = \sigma_{u0}^2 \begin{pmatrix} 1 \\ 1 \\ \vdots \\ 1 \end{pmatrix} + \sigma_{e0}^2 \begin{pmatrix} 1 \\ 0 \\ \vdots \\ 1 \end{pmatrix} + R \tag{2.11}$$

where R is a residual vector. The left-hand side of (2.11) is the response vector in the linear model and the right-hand side contains two explanatory variables, with coefficients $\sigma_{u0}^2, \sigma_{e0}^2$ which are to be estimated. The estimation involves an application of GLS using the estimated covariance matrix of $vec(\tilde{Y}\tilde{Y}^T)$, assuming Normality, namely $2(V^{-1} \otimes V^{-1})$ where $\otimes$ is the Kronecker product. The Normality assumption allows us to express this covariance matrix as a function of the random parameters. Even if the Normality assumption fails to hold, the resulting estimates are still consistent, although not fully efficient, but standard errors, estimated using the Normality assumption and, for example, confidence intervals will generally not be consistent. For certain variance component models alternative distributional assumptions have been studied, especially for discrete response models of the kind discussed in Chapter 4 (see for example Clayton and Kaldor, 1987) and maximum likelihood estimates obtained. For more general models, however, with several random coefficients, the assumption of multivariate Normality is a flexible one which allows a convenient parameterization for complex covariance structures at several levels. It is this assumption which forms the basis of the analyses in the remainder of the book, although in a later section we will look at a slight relaxation of this assumption where we fit the more general t-distribution for the level 1 residuals.

With the estimates obtained from applying GLS to (2.11) we return to (2.8) to obtain new estimates of the fixed effects and so alternate between the random and fixed parameter estimation until the procedure converges, that is the estimates for all the parameters do not change appreciably from one cycle to the next. Essentially the same procedure can be used for the more complicated models in the following chapters and is incorporated in the MLwiN program (Rasbash et al., 2000).

The maximum likelihood procedure produces biased estimates of the random parameters because it takes no account of the sampling variation of the fixed parameters. This may be important in small samples, and we can produce unbiased estimates by using a modification known as Restricted Maximum Likelihood (REML). The IGLS algorithm is readily modified to produce these restricted (RIGLS) estimates (Appendix 2.1).

Longford (1987) developed a procedure based upon a 'Fisher scoring' algorithm and Raudenbush (1994) shows that this is formally equivalent to IGLS. A variation on IGLS is Expected Generalized Least Squares (EGLS). This focuses interest on the fixed part parameters and uses the estimate of V obtained after the first iteration merely to obtain a consistent estimator of the fixed part coefficients without further iterations. A variant of this separates the level 1 variance from V as a parameter to be estimated iteratively along with the fixed part coefficients.

Another algorithm for obtaining ML or REML estimates is the EM algorithm or variants of it (Bryk and Raudenbush, 2002). This is outlined in Appendix 2.3 and has been incorporated into several software packages, partly because of its computational simplicity. In Chapter 4 we shall look at maximum likelihood and quasilikelihood procedures for generalized linear models with discrete responses.

2.5 Marginal models and Generalized Estimating Equations (GEE)

At this stage it is worth emphasizing the distinction between multilevel models, sometimes referred to in this context as 'subject specific' models, and so-called 'marginal' models such as the GEE model (Zeger et al., 1988; Liang et al., 1992). When dealing with hierarchical data these latter models typically start with a formulation for the covariance structure, for example but not necessarily based upon a multilevel structure, and aim to provide estimates with acceptable properties only for the fixed parameters in the model, treating the existence of any random parameters as a necessary 'nuisance'. More specifically, the estimation procedures used in marginal models are known to have useful asymptotic properties in the case where the exact form of the random structure is unknown.

If interest lies only in the fixed parameters, marginal models can be useful since they give unbiased estimates for these parameters. Even here, however, they may be inefficient if they utilize a covariance structure that is substantially incorrect. They are, however, generally more robust than multilevel models to serious misspecification of the covariance structure (Heagerty and Zeger, 2000). Fundamentally, however, marginal models address different research questions. From a multilevel perspective, the failure explicitly to model the covariance structure of complex data is to ignore information about variability that, potentially, is as important as knowledge of the average or fixed effects. Thus, in a repeated measures growth study, knowledge of how individual growth rates vary, possibly differentially according to say demographic factors, will be important data and in Chapter 5 we will show how such information can be used to provide efficient predictions in the case of human growth.

Also, when we discuss discrete response data multilevel models in Chapter 4 we will show how to obtain estimates for population or subpopulation means equivalent to those obtained from marginal models. In the case of Normal response linear multilevel models, GEE and multilevel models lead to the same fixed coefficient estimates. For a further discussion of the limitations of marginal models see the paper by Lindsey and Lambert (1998).

2.6   Residuals

In a single-level model such as (2.1) the usual estimate of the single residual term e_i is just $\tilde{y}_i$, the raw residual. In a multilevel model, however, we shall generally have several residuals at different levels. We can consider estimates for the individual residuals along the following lines.

Given the parameter estimates, consider predicting a specific residual, say u_{0j} in a 2-level variance components model. Specifically we require for each level 2 unit

$$\hat{u}_{0j} = E(u_{0j}|Y, \hat{\beta}, \hat{\Omega}) \tag{2.12}$$

We shall refer to these as estimated or predicted residuals or, using Bayesian terminology, as posterior residual estimates. If we ignore the sampling variation attached to the parameter estimates in (2.12) we have

$$\text{cov}(\tilde{y}_{ij}, u_{0j}) = \text{var}(u_{0j}) = \sigma_{u0}^2$$

$$\text{cov}(\tilde{y}_{ij}, e_{0ij}) = \sigma_{e0}^2 \tag{2.13}$$

$$\text{var}(\tilde{y}_{ij}) = \sigma_{u0}^2 + \sigma_{e0}^2$$

We may regard (2.12) as a linear regression of u_{0j} on the set of raw residuals $\{\tilde{y}_{ij}\}$ for the j-th level 2 unit and (2.13) defines the quantities required to estimate the regression coefficients and hence $\hat{u}_{0j}$. Details are given in Appendix 2.2. For the variance components model we obtain

$$\hat{u}_{0j} = \frac{n_j \sigma_u^2}{(n_j \sigma_u^2 + \sigma_{e0}^2)} \tilde{y}_j$$

$$\tilde{e}_{0ij} = \tilde{y}_{ij} - \hat{u}_{0j} \tag{2.14}$$

$$\tilde{y}_j = \left(\sum_i \tilde{y}_{ij}\right) \Big/ n_j$$

where n_j is the number of level 1 units in the j-th level 2 unit. Estimates are obtained by substituting sample values in (2.14). The residual estimates are not, unconditionally, unbiased but they are consistent. The factor multiplying the mean $(\tilde{y}_j)$ of the raw residuals for the j-th unit is often referred to as a 'shrinkage factor' since it is always less than or equal to one. As n_j increases this factor tends to one, and as the number of level 1 units in a level 2 unit decreases the 'shrinkage estimator' of u_{0j} becomes closer to zero. In many applications the higher level residuals are of interest in their own right and the increased shrinkage for a small level 2 unit can be regarded as expressing the relative lack of information in the unit so that the best estimate places the predicted residual close to the overall population value as given by the fixed part.

These residuals therefore can have two roles. Their basic interpretation is as random variables with a distribution whose parameter values tell us about the variation among the level 2 units, which provide efficient estimates for the fixed coefficients. A second interpretation is as individual estimates for each level 2 unit where we use the assumption that they belong to a population of units to predict their values. In particular, for units which have only a few level 1 units, we can obtain more precise estimates than if we were to ignore the population membership assumption and use only the information from those units. This becomes especially important for estimates of residuals for random coefficients, other than the intercept, where in the extreme case of only one level 1 unit in a level 2 unit we lack information to form an independent estimate. We shall illustrate this when we consider predictions based upon repeated measures growth models.

As in single-level models we can use the estimated residuals to help check on the assumptions of the model. The two particular assumptions that can be studied readily are the assumption of Normality and that the variances in the model are constant. Because the variance of the residual estimates depends in general on the values of the fixed coefficients it is common to standardize the residuals by dividing by the appropriate standard errors. The formulae for these are given in Appendix 2.2 where we refer to them as 'diagnostic' or 'unconditional' standard errors.

When the residuals at higher levels are of interest in their own right, we need to be able to provide interval estimates and significance tests as well as point estimates for them or functions of them. For these purposes we require estimates of the standard errors of the estimated residuals, where the sample estimate is viewed as a random realization from repeated sampling of the same higher level units whose unknown true values are of interest. The formulae for these 'conditional' or 'comparative' standard errors are also given in Appendix 2.2. For the estimates given in (2.12) the comparative variance is given by $\sigma_e^2 \sigma_u^2 (\sigma_e^2 + n_j \sigma_u^2)^{-1}$ and the diagnostic variance by $n_j \sigma_u^4 (\sigma_e^2 + n_j \sigma_u^2)^{-1}$.

The level 1 residuals are generally not of interest in their own right but are used rather for model checking, having first been standardized using the diagnostic standard errors.

2.7 The adequacy of Ordinary Least Squares estimates

In Appendix 2.1 we give the formulae for estimating the true standard errors for OLS estimates when a multilevel model applies. When the intra-unit correlations (variance partition coefficients) are small we can expect reasonably good agreement between the multilevel estimates and the simpler OLS ones. While it is difficult to give general guidelines about when OLS is an adequate alternative we can readily derive an explicit formula for the balanced 2-level variance components model using a simple regression equation with an intercept and a single explanatory variable

$$y_{ij} = \beta_0 + \beta_1 x_{ij} + u_j + e_{ij}$$

Write $\rho_y \rho_x$ for the intra-unit correlations for Y, X respectively and n for the number of level 1 units in the j-th level 2 unit. To obtain an estimate of the correct standard error for the estimate of β_1 we multiply the usual OLS estimate of the standard error by the quantity

$$\{1 + \rho_y \rho_x (n - 1)\}^{1/2}$$

Thus if there is exactly one level 1 unit per level 2 unit or either of the intra-unit correlations are zero, this expression is equal to 1.0 and the usual expression is correct. As n increases so the OLS estimator increasingly underestimates the true standard error. Thus with $\rho_y = \rho_x = 0.20$ and 76 level 1 units per level 2 unit the true standard error is, on average, twice the OLS estimate. Hence confidence intervals based on the OLS

Table 2.1   Variance components model applied to JSP data

Parameter	Estimate (s.e.)	OLS estimate (s.e.)
Fixed		
Intercept	13.90	13.80
8-year score	0.65 (0.025)	0.65 (0.026)
Random		
σ_{u0}^2 (between schools)	3.19 (1.0)	
σ_{e0}^2 (between students)	19.80 (1.1)	23.30 (1.2)
Variance partition coefficient	0.14	

estimate will be too short and significance tests will too often reject the null hypothesis. By designing a study where n is small we may be able to rely on OLS procedures to give adequate estimates for the fixed coefficients, but this would then not allow us to study any multilevel structures with adequate precision.

2.8   A 2-level example using longitudinal educational achievement data

We shall fit the simple 2-level variance components model (2.7) to the JSP data with the 11-year maths score as response and a single explanatory variable, the 8-year maths score, in addition to the constant term, equal to 1 and defining the intercept. The parameter values are displayed in Table 2.1 with the OLS estimates given for comparison.

Comparing the OLS with the multilevel estimates we see that the fixed coefficients are similar, but that there is a variance partition coefficient value of 0.14. The estimate of the standard error of the between-school variance is less than a third of the variance estimate, suggesting a value highly significantly different from zero.[2] This comparison, however, should be treated cautiously, since the variance estimate does not have a Normal distribution and the standard error is only estimated, although the size of the sample here will make the latter caveat less important. It is generally preferable to carry out a likelihood ratio test by estimating the 'deviance' for the current model and the model omitting the level 2 variance (see McCullagh and Nelder, 1989) and the next section will deal more generally with inference procedures. The deviances are, respectively, 4294.2 and 4357.3 with a difference of 63.1. This value would normally be referred to tables of the chi-squared distribution with one degree of freedom, and is highly significant. In the present case the null hypothesis of a zero variance is on the 'boundary' of the feasible parameter space; we do not envisage a negative variance. In this case the P-value to be used is half the one obtained from the tables of the chi-squared distribution (Self and Liang, 1987). Note that if we use the standard error estimate given in Table 2.1 to judge significance we obtain the corresponding value of $(3.19/1.0)^2 = 10.2$ which in this case is very much smaller than the likelihood ratio test statistic. Note also that if we use this test we would again use half the nominal P-value since only positive departures are possible.

---

[2] Since the intercept estimate depends upon the origins chosen for the other explanatory variables, its value generally has no substantive interest and we omit its standard error.

Table 2.2   Variance components model applied to JSP data with gender and social class

Parameter	Estimate (s.e.)	Estimate (s.e.)
Fixed		
Intercept	14.90	32.90
8-year score	0.64 (0.025)	
Gender (boys–girls)	−0.36 (0.34)	−0.39 (0.47)
Social class (non-manual–manual)	0.72 (0.39)	2.93 (0.51)
Random		
σ_{u0}^2 (between schools)	3.21 (1.0)	4.52 (1.5)
σ_{e0}^2 (between students)	19.60 (1.1)	37.20 (2.0)
Variance partition coefficient	0.14	0.11

We elaborate the model first by adding two more explanatory variables, gender and social class. The results are set out in the first column of Table 2.2.

The random parameter estimates are hardly changed, nor is the coefficient of the 8-year maths score. The gender difference is very small and in favour of the girls, but is far from the conventional 5% significance level. The social class difference favours the children of non-manual parents. When we are judging the fixed effects, a simple comparison of the estimate with its standard error is usually adequate. Because the model adjusts for the earlier maths score we can interpret the social class and gender differences in terms of the relative progress of girls versus boys or non-manual versus manual children. The second column in Table 2.2 shows the effects when 8-year maths score is removed from the model and the interpretation is now in terms of the actual differences found at 11 years. Note that the level 1 and level 2 variances are increased, reflecting the importance of the earlier score as a predictor, and the variance partition coefficient is slightly reduced. The social class difference is much larger, suggesting that most of the difference is that existing at 8 years with a somewhat greater progress made between 8 and 11 years by those in the non-manual social group. The gender difference remains small.

The 8-year score has been used as it stands, without centring it in any way. This is acceptable in the present case, although the strict interpretation of the intercept is the predicted score at age 11 of a child with an 8 year score of zero, which is outside the range of the observed values. If we were to measure the 8-year-score from its mean, the intercept would be interpreted as the predicted value at the mean 8-year-score. When we introduce random coefficients we shall see that this becomes an important consideration.

2.8.1   Checking for outlying units

Figure 2.7 is a plot of the estimated residuals against equivalent Normal scores (see section 2.11) and shows one school, identified as number 38, with the largest residual of 3.5. It is often useful to study the effect of omitting one or more units from an analysis to see what difference this makes to the parameter estimates and we shall say more about this below. For now we illustrate the effect of such units using school 38. Table 2.3 shows the parameter estimates associated with two different procedures.

In analysis A school 38 is simply omitted. The principal effect is to reduce the level 2 variance by about 14%, with little effect on the other parameters. In analysis B we have retained all the data in the analysis, but removed school 38 from the level 2 variation by fitting a separate Intercept term in the fixed part of the model. For the explanatory variable defining the level 2 variance we fit Z_0^* rather than Z_0, where

$$Z_0^* = \begin{cases} 0 & \text{if school 38} \\ 1 & \text{otherwise} \end{cases}$$

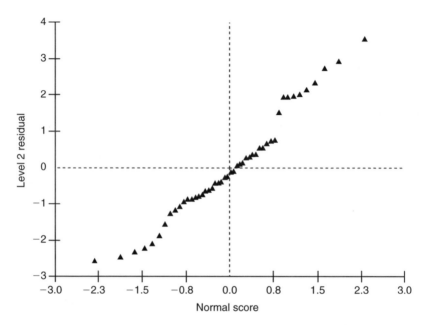

Figure 2.7 Standardized level 2 residuals by Normal equivalent scores for first column model in Table 2.2.

Table 2.3 As Table 2.2. Analysis A omitting school 38, analysis B fitting a constant for school 38

Parameter	Estimate (s.e.)	
	A	B
Fixed		
Intercept	14.5	14.7
8-year score	0.65 (0.026)	0.64 (0.025)
Gender (boys–girls)	−0.40 (0.34)	−0.37 (0.34)
Social class (non-manual–manual)	0.74 (0.39)	0.72 (0.38)
School 38		6.10 (1.5)
Random		
σ_{u0}^2 (between schools)	2.74 (0.9)	2.75 (0.9)
σ_{e0}^2 (between students)	19.60 (1.1)	19.60 (1.1)
Variance partition coefficient	0.12	0.12

and the Intercept fitted in the fixed part is simply $1 - Z_0^*$. The relatively small number of students, 9, in school 38 accounts for the fact that its shrunken residual mean of 3.5 is considerably less than the directly fitted mean of 6.1. Although it makes little difference to the parameter estimates in this example, in general it seems preferable to fit separate parameters for influential units and retain as much data as possible in the analysis.

2.8.2   Model checking using estimated residuals

We now check a particularly important assumption of the model by looking at the residuals. Figure 2.8 is a plot of the standardized level 1 residuals against the fixed part predicted value and Figure 2.9 is a plot of these residuals against their equivalent Normal scores. The latter plot is close enough to a straight line to give us some confidence that the Normality assumption is reasonable. Figure 2.8, however, shows the same pattern as Figure 2.1 of a decreasing variance with increasing 8-year score, so that the assumption of a constant level 1 variance is clearly untenable.

In Chapter 3 we shall be looking at ways of directly modelling such non-constant or complex level 1 variation. In the next section we look at general diagnostic procedures.

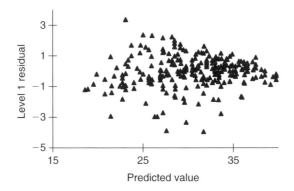

Figure 2.8   Standardized level 1 residuals by predicted values for Table 2.2.

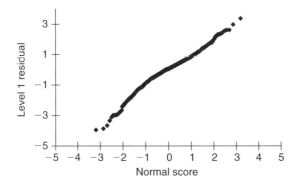

Figure 2.9   Standardized level 1 residuals by Normal equivalent scores for Table 2.2.

In the following section we look at nonlinear transformations of the response variable to produce a more constant variance.

2.9   General model diagnostics

Procedures for model exploration fall into two broad groups. The first of these is concerned with choosing the distributional assumptions, the set of variables for inclusion and the scales upon which they are measured. The choice of variables for inclusion, where external subject matter considerations are unavailable, is more complicated in multilevel models than in single-level models. In single-level linear models, stepwise procedures are available, which if used with care can produce usefully descriptive models. In multilevel models we need to consider random parameters as well as fixed coefficients and procedures for the partial automation of stepwise techniques do not seem to be available.

The second group of procedures is based upon analyses of the effects that individual units can have on parameter estimates. We have already explored some uses of the estimated residuals and how the omission or modification of units can change inferences. A related approach uses measures of influence, and is described in detail by Langford and Lewis (1998) and implemented in the MLwiN package (Rasbash et al., 2000). A somewhat different approach to model diagnosis is taken by Hodges (1998) who considers in particular the use of Markov Chain Monte Carlo (MCMC) chains.

The notion of 'influence' is that of introducing small changes to a model and assessing the effects of these on the resulting inferences. A central feature is 'leverage' which expresses the effect of deleting a particular unit on the parameter estimates. In a 2-level model Langford and Lewis use the set of values for the level 1 units in level 2 unit j given by

$$H_j = diag\left(V_j^{-0.5} X_j (X_j^T V_j^{-1} X_j)^{-1} X_j^T V_j^{-0.5} \right)$$

for the fixed parameters and an analogous formula for the random parameters. The larger the values the more potential they have for affecting the model parameter estimates.

Langford and Lewis also consider 'deletion' residuals, formed by deleting a particular unit, estimating from the reduced model and applying these parameter estimates to the complete dataset to calculate the residual for the deleted unit. This is approximately equivalent to fitting a dummy variable for the unit in question. They also discuss leverage values associated with random coefficients which may help the data analyst to assess the effect of a change in one or more of these. A detailed example explaining the use of these procedures is given by Lewis and Langford (2001).

When using diagnostic procedures some care is needed. Choosing to omit a unit or fit a dummy variable for it may perhaps best be regarded as part of a sensitivity analysis to see how robust the model is to varying the model assumptions. The use of formal significance tests when units have been detected on the basis that they are extreme also needs to be done carefully. A simple 'conservative' procedure that can be used when identifying outlying units is to multiply the significance levels (P-values) obtained for each test by the number of units at that level. Thus, from Table 2.3 the significance level associated with fitting a separate intercept for school 38 is 0.000023, and when multiplied by the number of schools (48) is 0.0011 which is still highly significant.

2.10   Higher level explanatory variables and compositional effects

We have already mentioned that from the point of view of estimating parameters, the explanatory variables can be defined or measured at any level. For substantive interpretations, however, explanatory variables measured at levels 2 or above often have particular interpretations. We illustrate some of these using the JSP dataset and forming the explanatory variable which is the mean 8-year-old maths score. This is often known as a 'compositional' variable since it measures an aspect of the composition of the school to which the individual student belongs. We are interested in whether the average 8-year score has an effect on the 11-year score, after having adjusted for the student's own 8-year score. For this analysis all the 8-year scores are measured about the sample mean value of 25.98 (Table 2.4). Analysis A adds the average school 8-year score. Its coefficient is very small and not significant. Analysis B uses the school-centred 8-year score. This is often advocated on the grounds that it is the difference between a student's score and the average score for that student's school which is likely to be the most relevant predictor of later achievement.

Bryk and Raudenbush (2002) give a detailed discussion of this issue for models where the compositional variable, as here, is a mean computed for all the students in the school, or more generally all the level 1 units in the relevant level 2 unit. Analyses A and B are, of course, formally equivalent and analysis A indicates directly that a simpler model omitting the school mean score is adequate. It is analysis C, as discussed below, which introduces a more complex model.

In fact, the mean score for students in a school is only one particular summary statistic describing the composition of the students. Another summary would be the spread of scores, measured for example by their standard deviation. We can also consider measures such as the proportions of high or low scoring students and in general any set of such measures. When using the average score we can also consider using the median or modal score rather than the mean. With any of these other measures we may wish to retain the deviation from the school mean as an explanatory variable, and

Table 2.4   Variance components model for JSP data with mean 8-year score measured about sample mean and centring about school mean

Parameter	Estimate (s.e.)		
	A	*B*	*C*
Fixed			
Intercept	31.50	31.50	31.70
8-year score	0.64 (0.025)		1.25 (0.26)
8-year score centred on school mean		0.64 (0.026)	
Gender (boys–girls)	−0.36 (0.34)	−0.36 (0.34)	−0.37 (0.34)
Social class (non-manual–manual)	0.72 (0.38)	0.72 (0.31)	0.79 (0.31)
School mean 8-year score	−0.01 (0.13)	0.63 (0.12)	−0.03 (0.12)
8-year score × school mean 8-year score			−0.02 (0.01)
Random			
σ_{u0}^2 (between schools)	3.21 (1.0)	3.21 (1.0)	3.13 (1.0)
σ_{e0}^2 (between students)	19.60 (1.1)	19.60 (1.0)	19.50 (1.1)
Variance partition coefficient	0.14	0.14	0.14

Table 2.5   Random coefficient model for JSP data

Parameter	Estimate (s.e.)
Fixed	
Intercept	31.70
8-year score	1.11 (0.35)
Gender (boys–girls)	−0.25 (0.32)
Social class (non-manual–manual)	0.96 (0.36)
School mean 8-year score	−0.04 (0.13)
8-year score × school mean 8-year score	−0.02 (0.01)
Random	
Level 2	
σ^2_{u0} (Intercept)	3.67 (1.03)
σ_{u01} (covariance)	−0.34 (0.09)
σ^2_{u1} (8-year score)	0.03 (0.01)
Level 1	
σ^2_{e0}	17.70 (1.0)

we could even consider introducing a more complex function of this, for example by adding higher order terms. There is here a fruitful area for further study.

Analysis C looks at the possibility of an interaction between student score and school mean and we do find a significant effect which we can interpret as follows. The higher the school mean 8-year score the lower the coefficient of the student's 8-year score. One implication of this is that for two relatively low scoring students at 8 years, the one in the school with a higher average is predicted to do better at 11 years. To study this further we now need to introduce a model with random coefficients where we explicitly allow each school's coefficient to vary randomly at level 2, as in Equation (2.6) (Table 2.5).

The addition of the 8-year score coefficient as a random variable at level 2 somewhat increases the social class difference and somewhat decreases the gender difference, but within their standard errors.

The level 1 variance is reduced and we have significant 'slope' variation at level 2; the likelihood ratio test criterion is 52.4 which is referred to chi-squared tables with 2 degrees of freedom and is highly significant. If we calculate the correlation between the intercept and slope at level 2 we obtain a value of −1.03! This sometimes happens as a result of sampling variation and implies that the population correlation is very high. We shall see in Chapter 3 we can constrain this correlation to be exactly −1.0 and thus admissible. Alternatively, by suitably elaborating the model or by carrying out certain transformations we can avoid this problem. For now, however, in order to illustrate what this means in the present data we can compute residuals for each school, for the slope and intercept. With these estimates we can then predict the 11-year score for any set of values of the explanatory variables. Figure 2.10 shows the predicted values for manual girls by 8-year score.

The predicted lines for the high scores at 8 years are very close together, separating as the 8-year score decreases. The slope residual is almost uncorrelated (−0.02) with the mean 8-year score and the compositional coefficient of mean 8-year score is little changed. We can add, therefore, to the previous compositional effect, the statement that some schools are differentially 'effective' for pupils with low 8-year scores, with little difference for high 8-year scores. In Chapter 3 we shall continue to analyse this

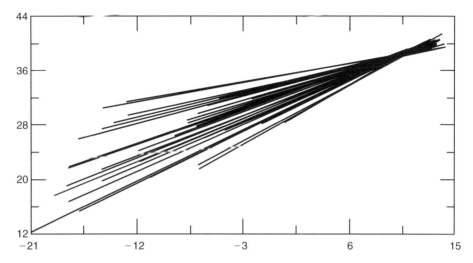

Figure 2.10 Predicted 11-year score by 8-year score for JSP schools.

dataset and show how further elaboration of the variance structure of the model leads
to certain simplifications of interpretation.

2.11 Transforming to Normality

For single-level data there is a considerable literature on ways of transforming measurements to satisfy the standard model assumption of Normality with constant variance
(see Box and Cox, 1964). Cole and Green (1992), for example, describe a procedure
for smoothly modelling the mean, variance and skewness of measurements repeated
over time and using the estimated smoothed relationships to make the appropriate
transformation to Normality. We shall use a general purpose procedure designed to produce approximately Normally distributed residuals and study its effect on our example
analyses.

   If we look at Figure 2.9 we can imagine 'stretching' the Y axis to make the plot
follow a straight line. Of course, we cannot do this directly since the residuals are estimates rather than observed measurements, but we can do this for the original response
measurements with the expectation that this will give us residuals which are more
nearly Normally distributed, and also with a more constant variance. To do this we
rank all the response measurements and assign to each one the equivalent value from a
standard Normal distribution that cuts off the same proportion of the population as does
the ranked observation. Thus, for example, if we have 99 measurements, the smallest
value is a non-parametric estimate of the first percentile of the population distribution;
the value such that just 1% lie below this value. We would then assign this observation the equivalent Normal score of -2.33 which is the lower 1% point of the $N(0, 1)$
distribution. This procedure will often be justified, for example with educational test
scores, where the measurement scales themselves are essentially arbitrary, but in other
cases, for example with physical measurements, the original scale is meaningful and in
such cases we would usually prefer to use the variance modelling methods described
in Chapter 3.

Table 2.6 Variance components model applied to JSP data with gender and social class and Normalized 11-year and 8-year scores

Parameter	Estimate (s.e.)
Fixed	
Intercept	0.129
8-year score	0.668 (0.026)
Gender (boys–girls)	−0.048 (0.050)
Social class (non-manual–manual)	0.138 (0.057)
Random	
σ^2_{u0} (between schools)	0.080 (0.023)
σ^2_{e0} (between students)	0.422 (0.023)
Variance partition coefficient	0.16

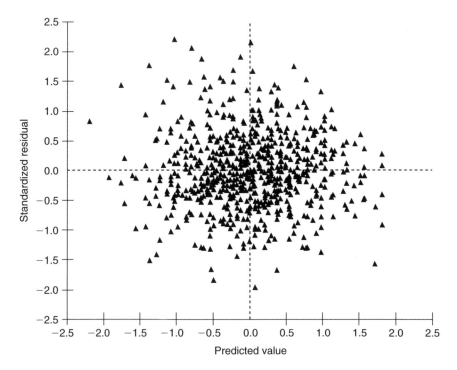

Figure 2.11 Standardized level 1 residuals by predicted value for Table 2.6.

We now repeat the analysis in Table 2.2 for our transformed data, where we have also transformed the 8-year score in similar fashion. Table 2.6 shows the results. The inferences we would make from Table 2.6 are similar to the earlier ones, except that the social class coefficient is now more significant and the intra-school correlation has increased slightly. Figure 2.11, however, shows that the variance of the level 1 residuals is now more nearly constant, although not entirely so, and Figure 2.12 shows that the residuals follow a Normal distribution more closely also. In Chapter 3 we shall look at further modelling of the variance for these transformed data.

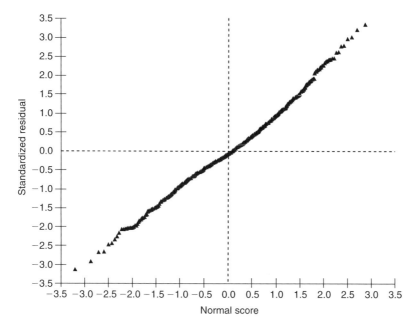

Figure 2.12   Standardized level 1 residuals by Normal equivalent scores for Table 2.6.

2.12   Hypothesis testing and confidence intervals

In this section we deal with large sample procedures for constructing interval estimates for parameters or linear functions of parameters and for hypothesis testing. Hypothesis tests are used sparingly throughout this book, since the usual form of a null hypothesis, that a parameter value or a function of parameter values is zero, is usually implausible and also relatively uninteresting. Moreover, with large enough samples a null hypothesis will almost certainly be rejected. The exception to this is where we are interested in whether a difference is positive or negative, and this is discussed in the section on residuals below. Confidence intervals emphasize the uncertainty surrounding the parameter estimates and the importance of their substantive significance.

2.12.1   Fixed parameters

In the analyses of section 2.11 we presented parameter estimates for the fixed part parameters together with their standard errors. These are adequate for hypothesis testing or confidence interval construction separately for each parameter. In many cases, however, we are interested in combinations of parameters. For hypothesis testing, this most often arises for grouped or categorized explanatory variables where n group effects are defined in terms $n - 1$ of dummy variable contrasts and we wish simultaneously to test whether these contrasts are zero. In the case of the analysis in Table 2.2 we may be interested in the hypothesis that the gender and social class effects, taken jointly, are zero. We may also be interested in providing a pair of confidence intervals for the parameter estimates. We proceed as follows.

Define a $(r \times p)$ contrast matrix C. This is used to form linearly independent functions of the p fixed parameters in the model of the form $f = C\beta$, so that each row of C defines a particular linear function. Parameters which are not involved have the corresponding elements set to zero. Suppose we wish to test the hypothesis in Table 2.2 that the gender and social class coefficients are jointly zero. We define

$$C = \begin{pmatrix} 0 & 0 & 1 & 0 \\ 0 & 0 & 0 & 1 \end{pmatrix}, \quad f = \begin{pmatrix} \beta_2 \\ \beta_3 \end{pmatrix}$$

and the general null hypothesis is

$$H_0 : f = k, \quad k = \{0\} \text{ here}$$

We form

$$R = (\hat{f} - k)^T [C(X^T \hat{V}^{-1} X)^{-1} C^T]^{-1} (\hat{f} - k)$$

$$\hat{f} = C\hat{\beta} \tag{2.15}$$

If the null hypothesis is true this is distributed as approximately χ^2 with r degrees of freedom. This is often known as a 'Wald test'. Note that the term $(X^T \hat{V}^{-1} X)^{-1}$ is the estimated covariance matrix of the fixed coefficients.

If we find a statistically significant result we may wish to explore which particular linear combinations of the coefficients involved are significantly different from zero. The common instance of this is where we find that n groups differ and we wish to carry out all possible pairwise comparisons. A simultaneous comparisons procedure which maintains the overall type I error at the specified level involves carrying out the above procedure with either a subset of the rows of C or a set of (less than r) linearly independent contrasts. The value of R obtained is then judged against the critical values of the χ^2 distribution with r degrees of freedom.

We can also obtain a $(100 - \alpha)\%$ confidence region for the parameters by setting $\hat{R}$ equal to $\alpha\%$ the tail region of the χ^2 distribution with r degrees of freedom in the expression

$$\hat{R} = (f - \hat{f})^T [C(X^T \hat{V}^{-1} X)^{-1} C^T]^{-1} (f - \hat{f})$$

This yields a quadratic function of the estimated coefficients, giving an r-dimensional ellipsoidal region. For Table 2.2 we obtain the following results.

The null hypothesis test gives a value for χ^2 on 2 degrees of freedom of 4.51 with a corresponding P-value of 0.10. The 95% confidence region is the ellipse

$$8.3(\beta_1 + 0.36)^2 + 0.22(\beta_1 + 0.36)(\beta_2 - 0.72) + 6.7(\beta_2 - 0.72)^2 = 5.99$$

where the subscripts $(1, 2)$ refer to gender and social class respectively and 5.99 is the 5% point of the χ_2^2 distribution. Figure 2.13 displays this region, which contains the point $(0, 0)$ so that the null hypothesis that $\beta_1 = \beta_2 = 0$ is not rejected at the 5% level.

In some situations we may be interested in separate confidence intervals for all possible linear functions involving a subset of q parameters or q linearly independent functions of the parameters, while maintaining a fixed probability that all the intervals include the population value of these functions of the parameters. As before, this

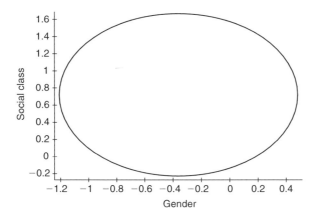

Figure 2.13 95% confidence region for coefficients of social class and gender.

may arise when we have an explanatory variable with several categories and we are interested in intervals for sets of contrasts. For a $(100 - \alpha)\%$ interval write C_i for the i-th row of C, then a simultaneous $(100 - \alpha)\%$ interval for $C_i \beta$, for all C_i is given by

$$(C_i \hat{\beta} - d_i, \ C_i \hat{\beta} + d_i)$$

where

$$d_i = [C_i (X^T \hat{V}^{-1} X)^{-1} C_i^T \ \chi^2_{q,(\alpha)}]^{0.5}$$

where $\chi^2_{q,(\alpha)}$ is the $\alpha\%$ point of the χ^2_q distribution. For model A of Table 2.2 we obtain the following 95% intervals for the coefficients of gender and social class, first the separate intervals then the simultaneous ones which are some 25% wider.

$$\begin{pmatrix} -0.36 \pm 0.66 \\ 0.72 \pm 0.76 \end{pmatrix}, \quad \begin{pmatrix} -0.36 \pm 0.83 \\ 0.72 \pm 0.94 \end{pmatrix}$$

We can also use the likelihood ratio test criterion for testing hypotheses about the fixed parameters, although generally the results will be similar. The difference arises because the random parameter estimates used in (2.15) are those obtained for the full model rather than those under the null hypothesis assumption, although this modification can easily be made. For example the likelihood ratio test for gender and social class yields a value of 5.5 compared with the above value of 4.5. We shall discuss the likelihood ratio test in the next section dealing with the random parameters.

2.12.2 Random parameters

In very large samples it is possible to use the same procedures for hypothesis testing and confidence intervals as for the fixed parameters. Generally, however, procedures based upon the likelihood statistic are preferable. To test a null hypothesis H_0 against an alternative H_1, involving the fitting of additional parameters we form the log likelihood ratio or deviance statistic

$$D_{01} = -2 \log_e (\lambda_0 / \lambda_1) \tag{2.16}$$

where λ_0, λ_1 are the likelihoods for the null and alternative hypotheses and this is referred to tables of the chi squared distribution with degrees of freedom equal to the difference (q) in the number of parameters fitted under the two models. We have already quoted this statistic for testing the level 2 variance in Table 2.1, with the caveat needed when the null model involves a 'boundary' value for the parameters. A similar caveat applies to the calculation of confidence regions or variance parameters not significantly different from zero.

We can also use (2.16) as the basis for constructing a $(100 - \alpha)\%$ confidence region for the additional parameters. If D_{01} is set to the value of the $\alpha\%$ point of the χ^2 distribution with q degrees of freedom, then a region is constructed to satisfy (2.16), using a suitable search procedure. This is a computationally intensive task, however, since all the parameter estimates are recomputed for each search point.

An alternative is to use the 'profile likelihood' (McCullagh and Nelder, 1989). In this case the likelihood is computed for a suitable region containing values of the random parameters of interest, for fixed values of the remaining random parameters.

Unlike ML estimates we cannot use REML (RIGLS) estimates to compute a likelihood that can then be used in a general way to compare models, although the REML likelihood can be used if we are just making comparisons between models that have the same set of fixed effects but different random parameters. Thus, for model exploration and hypothesis testing it is recommended that ML (IGLS) estimation should be used, with a final REML estimation if this is required. When we look at MCMC estimation in section 2.13 we shall see how exact interval estimates can be constructed and in Chapter 3 we shall see how this can be done using the bootstrap.

2.12.3 Hypothesis testing for non-nested models

So far we have considered whether we should accept a sub-model from a model with extra parameters, for example whether a variance term is significantly different from zero. In some cases, however, we may wish to compare two models where one is not a strict subset of the other; such models are referred to as 'non-nested'. Consider the simple variance components model for the JSP data from Table 2.1 where we use transformations of the 11-year and 8-year scores so that these now have standard Normal distributions.

Table 2.7 gives the results of fitting two models: one (model A) uses the Normalized 8-year score as single predictor and the other (model B) uses a logarithmic transformation together with its square. For model A the addition of a squared term changes the likelihood by a very small amount, but for model B we obtain a significant coefficient for the squared term so this remains in this model.

We see that the likelihood statistics differ, but we cannot in this case simply take the difference and use this to judge between models. Where the models are not nested in this way, we can use the Akaike Information Criterion (AIC), as a tool to search for the best fitting model. For each model the AIC is simply $l + 2p$, $(l = -2\log_e(\lambda))$, where p is the number of parameters fitted in the model and the model with the smallest AIC is chosen as the one which fits best. Note that there is no probabilistic interpretation here that would allow us to say whether one model is significantly better than the other; the AIC is simply an index for model comparison. We are fitting respectively 4 and 5 parameters so that in the present case the AIC for the log-transformed model is $1504.0 + 2 * 5 = 1514.0$ and for the untransformed model it is $1505.7 + 2 * 4 = 1513.7$. We see therefore that, despite the larger number of parameters in model B,

Table 2.7   Variance components model applied to Normalized JSP data

Parameter	Model A	Model B
Fixed		
Intercept	0.01	−1.75
8-year score (x)	0.680 (0.026)	
$\log_e (x + 3.5)$		0.430 (0.21)
$[\log_e (x + 3.5)]^2$		0.800 (0.10)
Random		
σ^2_{u0} (between schools)	0.079 (0.023)	0.082 (0.024)
σ^2_{e0} (between students)	0.427 (0.023)	0.425 (0.023)
−2 (log-likelihood)	1505.7	1504.0

model A has the smaller AIC, although there is a negligible difference between the values. Lindsey (1999) gives a discussion of the use of this criterion.

An alternative to the AIC is the Bayesian Information Criterion (BIC) which is computed as $l + \log_e(N^*)p$, $(l = -2\log_e(\lambda))$. The term N^* is the effective sample size. In a multilevel model it is not clear what the effective sample size should be, and the total number of higher level units is often used as an approximation. Raftery (1995) provides a discussion. For both criteria we note that the comparisons must be based on the same sample of units.

2.12.4   Inferences for residual estimates

In our JSP variance components analysis we estimated level 2 residuals, one for each school. In studies of school effectiveness, one requirement is sometimes to try to identify schools with residuals which are substantially different. From a significance testing standpoint, we will often be interested in the null hypothesis that school A has a smaller residual than school B against the alternative that the residual for school A is larger than that for school B (ignoring the vanishingly small probability that they are equal). In the case when a standard significance test accepts the alternative hypothesis (at a chosen level, say $\alpha\%$) of some difference against the null hypothesis of no difference, this is equivalent to accepting one of the alternatives ($A > B, A < B$) at the same level of significance and we shall use this interpretation. Alternatively we may construct a $(100 - \alpha)\%$ confidence interval for the difference between two schools' means and see whether it contains zero, is less than or greater than zero.

Where we can identify two particular schools then it is straightforward, using the results of Appendix 2.1, to construct a confidence interval for their difference or carry out a significance test. Often, however, the results are made available to a number of individuals, who are each interested in comparing their own schools of interest. This may occur, for example, where policy makers wish to select a few schools within a small geographical area for comparison, out of a much larger study. In the following discussion, we suppose that individuals wish to compare only pairs of schools, although the procedure can be extended to multiple comparisons of three or more residuals. Further details are given by Goldstein and Healy (1995).

Consider the JSP data where we have 48 estimated residuals together with their comparative standard errors. Since the sample size is fairly large, we can also assume that these estimates are uncorrelated.

Firstly, we order the residuals from smallest to largest. We construct an interval about each residual so that the criterion for judging statistical significance at the $(1 - \alpha)\%$ level for any pair of residuals is whether their confidence intervals overlap. For example, if we consider a pair of residuals with a common standard error (se), and assuming Normality, the confidence interval width for judging a difference significant at the 5% level is given by $\pm 1.39(\text{se})$.

The general procedure defines a set of confidence intervals for each residual as

$$\hat{u}_j \pm c(\text{se})_j \tag{2.17}$$

For each possible pair of intervals (2.17), there is a significance level associated with the overlap criterion, and the value c is determined so that the average, over all possible pairs, is $\alpha\%$. A search procedure can be devised to determine c. When the ratios of the standard errors do not vary appreciably, say by not more than 2:1, the value 1.4 can be used for c. As this ratio increases so does the value of c. In the present case all but 2 of these ratios are greater than 2 and we have used the common value of 1.4.

The results are presented as the 'caterpillar plot' in Figure 2.14. As is clear, apart from some of the extreme intervals, each interval overlaps with most of the other intervals. If we wished the basic comparison to take place among triplets of schools, with simultaneous confidence intervals, then using the results of section 2.12.1 we replace the Normal upper 2.5% value of 1.96 by $\sqrt{\chi^2_{2,(0.05)}} = 2.45$, since we can assume that the residuals are approximately independently distributed. This will give a similar display but with intervals 25% wider. In reality the complete set of schools typically will be compared in overlapping subsets of different sizes, and a value for c can be determined by averaging over all such possibilities.

In other situations we may be interested simply in judging each school against the average for all schools. In this case conventional $\alpha\%$ intervals for the residuals can be constructed and inferences based upon whether they include zero (the mean of the residuals). If several such comparisons are made simultaneously then an adjustment to the intervals is required, noting again that we can usually regard the residuals as approximately independently distributed.

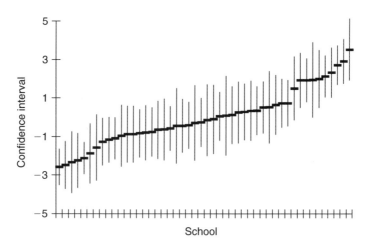

Figure 2.14   Simultaneous confidence intervals for JSP school residuals.

Presentations such as that in Figure 2.14 are useful for conveying the inherent uncertainty associated with estimates for individual level 2 (or higher) units, where the number of level 1 units per higher level unit is not large. This uncertainty in turn places inherent limitations upon such comparisons. See Goldstein and Spiegelhalter (1996) and Yang et al. (1999) for a detailed discussion of these issues in health and education.

2.13   Bayesian estimation using Markov Chain Monte Carlo (MCMC)

We now look at Bayesian models for multilevel data using MCMC methods. These incorporate prior distribution assumptions and, based upon successively sampling from posterior distributions of the model parameters, yield a 'chain' which is then used for making point and interval estimates. The two most common procedures in use are 'Gibbs sampling' and 'Metropolis Hastings sampling'. We shall illustrate how these work for some basic models, with examples. For more details about MCMC methodology see for example Gilks et al. (1996). In later chapters we will describe the use of MCMC methods as the basis for fitting more complex models.

All MCMC algorithms are iterative and at each iteration they are designed to produce a sample from the joint posterior (multivariate) distribution of the components of the model. These components will be variates, regression coefficients, covariance matrices, residuals, etc. After a suitable number of iterations, we obtain a sample of values from the distribution of any component which we can then use to derive any desired distribution characteristic such as the mean, covariance matrix, etc.

We outline the procedures for a general 2-level variance components model

$$y_{ij} = (X\beta)_{ij} + u_j + e_{ij}, \quad \text{var}(e_{ij}) = \sigma_e^2, \quad \text{var}(u_j) = \sigma_u^2 \qquad (2.18)$$

In the Bayesian formulation of this model we combine *prior* information about the fixed and random parameters, with the likelihood based on the data. These parameters are regarded as random variables described by probability distributions, and the prior information for a parameter is incorporated into the model via a *prior distribution*. After fitting the model, a distribution is produced for the above parameters, known as the *posterior distribution*. Formally we write for the posterior distribution

$$p(\theta|y) \propto L(y; \theta)p(\theta)$$

where θ represents the unknown parameters, y represents the observed responses and L, p respectively are the likelihood and the prior distribution for θ where we assume independent prior distributions for each component. Given a chain from the posterior distribution we can construct useful summaries, either in the form of point estimates such as the mean or mode, or for quantiles such as the upper and lower $(\alpha/2)\%$ points of the distribution for a single parameter which results in an interval estimate. The Bayesian interpretation therefore differs from the classical frequentist interpretation where a confidence interval is required to include the single true population mean with probability $(100 - \alpha)\%$. We shall not make a rigid distinction between these interpretations when describing results which typically can be viewed from either standpoint.

The topic of the choice of priors, and in particular whether and when to use informative priors in a Bayesian fashion, is a vast one. A useful introduction is Draper (2002), and from time to time we will refer to the use of informative priors, for example when discussing meta analysis in Chapter 5. Our default assumption, however, is that diffuse priors are used, in an attempt to base inference solely on the data and the results of doing this will typically be similar to the results from IGLS and RIGLS estimation. At the end of this chapter we shall see how MCMC methods can be adapted to obtain (approximate) maximum likelihood estimates. Two common methods for producing chains sampled from the joint posterior distribution of all the parameters are Gibbs sampling and Metropolis Hastings (MH) sampling and we consider these in turn.

2.13.1 Gibbs sampling

MCMC estimation is iterative and at each iteration produces a set of 'current' parameter values, and after convergence the sequence of these, under general conditions, can be considered as a serially correlated random draw from the joint posterior distribution of the parameters. It proceeds, at each iteration, by considering each component in turn and generating a random sample from the distribution of that component assuming the current values of the remaining components. The detailed steps are given in Appendix 2.4 where the choice of prior distributions is also considered. Here we outline the steps briefly. It is assumed that we have some starting values, derived say from a preliminary IGLS estimation.

In Gibbs sampling for the variance components model (2.8) the following steps occur at iteration t. Sampling of each set of parameters is conditional on current values of the others, the priors and the data.

- **Step 1**
  Sample a new set of fixed effects (β).
- **Step 2**
  Sample a new set of residuals $\{u_j\}$.
- **Step 3**
  Sample a new level 2 variance.
- **Step 4**
  Sample a new level 1 variance.
- **Step 5**
  Compute the level 1 residuals by subtraction.

The first few values from the MCMC chain will usually be discarded, the 'burn in' phase, and inference will be based upon the set obtained after the chain has converged to the joint posterior distribution. The chain of parameters is treated as a (serially correlated) random sample from the full joint posterior distribution of the parameters. Typically, the mean of the chain for each parameter is chosen as a point estimate and the standard deviation as an estimate of spread, equivalent to the standard error. The modal values, which correspond more closely to the maximum likelihood estimates when diffuse or uninformative priors are used, can also be useful. Distribution quantiles can be computed to provide interval estimates, and these can be obtained from the chain either by ranking the chain values and selecting the values corresponding to the desired quantiles, or by estimating the probability density distribution of the chain values by a suitable non-parametric smoothing method such as kernel density estimation (Silverman, 1986; Abdous and Berlinet, 1998). We can also form derived chains for functions of parameters. For example, the variance partition coefficient may be of interest and we can form a chain for this simply by calculating the appropriate ratio of the level 2 variance to the sum of the level 2 and level 1 variances at each iteration and storing these. We shall discuss these procedures in more detail in the examples below.

2.13.2 Metropolis Hastings (MH) sampling

Gibbs sampling can be used whenever we can write down the conditional distributions for each parameter or group of parameters in turn. When this cannot be done easily alternative approaches can be used. Here we describe briefly Metropolis Hastings sampling; for details of the alternative 'adaptive rejection sampling' see Gilks and Wild (1992).

The most important case when we need to use MH sampling is with discrete response, or multilevel generalized linear, models (see Chapter 4). The conditional distributions for the fixed parameters and the residuals involve exponential functions and these are not easy to sample from, so that another kind of sampling such as MH becomes necessary.

For a set of parameters at any given stage of the iterative cycles, MH proceeds by forming a *proposal distribution* for the parameters and sampling a new set of parameters from this distribution. A rule for accepting this new set or rejecting it is applied and if the proposed set is rejected we remain with the current values. This is repeated for each set of parameters in turn and, like Gibbs, chains for all the parameters are produced. Ideally we would like to have a proposal distribution that minimizes the number of iterations for a given accuracy. One simple approach to creating a proposal distribution is to sample from a multivariate normal distribution. Where this distribution is conditioned on the current parameter values it is known as a *random walk Metropolis sampler*. Further details of MH sampling are given in Appendix 2.4.

In practice for sampling the variance and covariance parameters, for example in generalized linear models, this can be carried out using Gibbs and this gives an efficient hybrid algorithm (see Browne and Draper, 2000) which is a mixture of MH and Gibbs sampling steps.

2.13.3 Convergence of MCMC chains

For the IGLS and RIGLS algorithms we can judge convergence by studying the relative change in successive parameter values for each parameter so that we can be assured that the estimate is accurate to a satisfactory number of significant digits. When this change has reached a satisfactory small value (e.g. 0.01) convergence is achieved and we can be fairly certain that, except in exceptional cases where a model is badly misspecified, further iterations will not alter these values. With MCMC chains we likewise wish to judge when we have sufficient accuracy for the mean or mode estimate of each parameter. Here, however, because of the stochastic nature of the chain we cannot be certain that we have reached a final set of values that would not change with further iterations.

In fact there are two ways of judging convergence for chain means. We can use the standard deviation of the chain parameter values (which is the estimated standard error of the parameter) divided by the square root of the number of chain values scaled by a suitable factor to take account of the non-independence of successive chain values, to give an estimate of the (Monte Carlo) standard error of the parameter mean (see below). If we divide this by the mean itself we can use this to make a judgement when it reaches a suitably small value such as 0.01. For parameters which have values close to zero, however, this has little utility and a diagnostic based upon the number of significant digits accuracy is required. Such a diagnostic would provide an estimate of the number of significant digits for a given chain with a specified accuracy for the result. To avoid specifying fractional significant digits this is better expressed in terms of the chain length needed to obtain, say, k significant digits with a specified relative accuracy. We shall look at one such diagnostic (Draper, 2002) in our example below.

One of the features of MCMC estimation is that the chain provides an estimate of the probability distribution for the parameter or for a set of parameters, and hence can be used to give interval estimates. From a frequentist viewpoint these can be interpreted as confidence intervals and from a Bayesian perspective as credible intervals for the parameter value. The simplest method for computing these involves ranking the chain values and reading off the quantiles which cut the observed cumulative distribution at the required percentages, for example 2.5% and 97.5%. Alternatively a smoothed estimated kernel density function can be used. As with the mean we require an estimate of the accuracy of such quantile estimates and a commonly used one (Raftery and Lewis, 1992) estimates the chain length needed for a specified accuracy for a particular quantile.

One of the major drawbacks to MCMC methods is the amount of real time that can be taken for satisfactory convergence. As we have already noted, the successive values generated in an MCMC chain are not independent since each one in turn is conditional (partly) on the previous one. If the correlation between successive values is very high then each additional value generates only a little extra information so that the chain will have to run for a comparatively long time. We can calculate the sample size based upon independent successive values that would give the same estimate for the standard error of the parameter mean. Thus, for example, if partial correlations of the second order and above are assumed to be negligible and the first order serial correlation is small, say less than 0.2, the variance of the parameter mean is $\mathrm{var}(\bar{x}_p) \sim (\sigma_p^2/n)[1 + \rho_1(2 - \rho_1)/(1 - \rho_1)]$ so that the equivalent sample size is $n(1 - \rho_1)/[1 + \rho_1(1 - \rho_1)]$. For larger first order correlations further terms $(\sigma_p^2/n)[\rho_1^2/(1 - \rho_1)]$, $(\sigma_p^2/n)[\rho_1^3/(1 - \rho_1)]$, etc. should be added to the approximate expression for $\mathrm{var}(\bar{x}_p)$. It is therefore useful to monitor this autocorrelation as we shall see in the example. Likewise, especially with MH sampling, the 'trace' of the chain of values should sweep across the full distribution of the parameter without staying too long at any one set of values; such traces are said to 'mix well' and we will be looking at such traces in an example. It may often be important to experiment with setting different values for the MH proposals to achieve 'good' traces.

The topic of MCMC chain diagnostics is an ongoing area of research and particular choices are not guaranteed to perform adequately in all circumstances.

2.13.4   Making inferences

We have explained how the chain values can be used to provide interval estimates for each parameter or a function of parameters. We can also study, for example, bivariate regions, especially if we have a pair of parameters whose joint posterior distribution can be approximated by a bivariate normal distribution. In such a case we can apply the methods of section 2.13.1.

We have seen how to obtain an index, the AIC, for comparing models which can be used when these were non-nested. For MCMC models we can use a similar criterion, a generalization of the AIC, which likewise allows comparison of models with different sets of parameters for a given set of responses and priors. This is known as the deviance information criterion (DIC) (Spiegelhalter et al., 2002) and is computed as follows.

At each iteration of the chain we compute the current value of -2 log-likelihood for the model being fitted using the current chain values, say D_i and write $\bar{D}$ for the mean value of this. Upon convergence we compute the deviance based upon the mean parameter values from the chain, say D. We then form $\bar{D} - D$ which is an estimate of the 'effective number of parameters' (p_D) in the model and the DIC statistic is simply $D + 2p_D$. In single-level models with diffuse priors the DIC value will be very close to the AIC value and the effective number of parameters very close to the actual number of parameters in the model. For multilevel models, however, this will no longer hold. Instead, the calculation of the effective number of parameters includes the residuals which are treated as parameters to be estimated, but because they are constrained by the distributional assumptions made about them, their effective number is less than their actual number, as we shall see in our example.

2.13.5 An example

We shall illustrate MCMC estimation here with a simple example. In later chapters when we deal with specific models we will often be applying MCMC as well as (R)IGLS methods and where appropriate will describe any new types of MCMC steps.

We use the JSP data and the variance components model of Table 2.4 and we shall fit a model using Gibbs sampling with a burn in of 500 iterations and a chain of 5000 iterations. We will use diffuse priors, looking at both the uniform and inverse gamma priors for the variance parameters. The results, together with the IGLS (ML) estimates, are given in Tables 2.8–2.10.

Table 2.8 IGLS and MCMC estimates for JSP data with Normalized scores with alternative (diffuse) priors for variances (standard errors are in parentheses)

	IGLS	Gibbs	
		Uniform priors	Inverse gamma priors
Fixed			
Intercept	0.129	0.130	0.130
0-year score	0.668 (0.026)	0.669 (0.027)	0.669 (0.027)
Gender (boys–girls)	−0.048 (0.050)	−0.047 (0.050)	−0.047 (0.050)
Social class			
(non-manual–manual)	0.138 (0.056)	−0.139 (0.057)	−0.138 (0.057)
Random			
Level 2			
σ_{u0}^2	0.080 (0.023)	0.093 (0.028)	0.086 (0.025)
Level 1			
σ_{e0}^2	0.422 (0.023)	0.426 (0.024)	0.426 (0.023)
Effective number			
of parameters (p_D)		38.7	38.1
DIC		1481.3	1481.1
Estimation time*	0.8 secs	16 secs	16 secs

*On a 600 MHz Pentium running Windows 2000 using MLwiN. MCMC burn in of 500 and main chain of 5000.

Table 2.9 MCMC estimates (selected parameters) for model in Table 2.8 with alternative estimates for location and quantile estimates: uniform priors for variance parameters

	Mean	Median	Mode	Standard deviation	2.5%, 97.5% quantiles
8-year score	0.669 (0.0004)	0.669	0.669	0.027	0.617, 0.721
Gender (boys–girls)	−0.047 (0.0007)	−0.047	−0.047	0.050	−0.146, 0.050
Random					
Level 2: σ_{u0}^2	0.093 (0.0006)	0.089	0.085	0.028	0.051, 0.159
Level 1: σ_{e0}^2	0.426 (0.0004)	0.426	0.415	0.024	0.382, 0.476

Table 2.10 MCMC estimates (selected parameters) for model in Table 2.8 with alternative estimates for location and quantile estimates: inverse gamma (0.001, 0.001) priors for variance parameters

	Mean	Median	Mode	Standard deviation	2.5%, 97.5% quantiles
8-year score	0.669 (0.0004)	0.669	0.669	0.027	0.617, 0.721
Gender (boys–girls)	−0.047 (0.0007)	−0.047	−0.047	0.050	−0.146, −0.050
Random					
Level 2: σ_{u0}^2	0.086 (0.0006)	0.082	0.078	0.025	0.047, 0.147
Level 1: σ_{e0}^2	0.426 (0.0004)	0.425	0.424	0.023	0.381, 0.474

Several features emerge from these comparisons. We see that the P_D statistic is about 38; the fixed coefficients and variances give 6 standard parameters; the 48 schools are described by 32 effective parameters. The main difference between the ML and MCMC estimates is in the level 2 variances and here the gamma priors produce results closer to the ML estimates and this will often be found (see also Browne, 1998). With just 48 schools we might expect that the choice of priors would have some effect on any parameters which depend on this number. By contrast the other parameters are based upon the total sample size so that the effect of a particular diffuse prior is very small. The other thing to notice is that the use of the standard error and a normality assumption in order to calculate an interval estimate may be misleading. Thus, for example for the level 2 variance this would give a symmetrical 95% interval of (0.036, 0.136) as opposed to the one estimated from the chain of (0.047, 0.147) for the gamma prior. We also note that the MCMC modal estimates are closer to the ML estimates, especially with the gamma prior assumption.

We also see that the means are fairly accurately estimated to judge from the MCSE values. Running the chain for longer will change the values slightly and we now look at some formal diagnostics.

Figure 2.15, produced by MLwiN, is for the level 2 intercept variance. It suggests that a sample size of 5000 is adequate for estimates of the 2.5% and 97.5% quantiles, but that a much longer run seems to be required for the mean to be accurate to two significant figures; however, running for 25 000 iterations does not change any values appreciably. For the quantiles this sample size determines with 95% probability ($s = 0.95$) that the (95%) interval between the quantiles should vary by no more than $100 * 2 * 0.005 = 1$ percentage point ($r = 0.005$). These diagnostics, however, are not always reliable and it is often important to continue running a chain to see how the estimates behave.

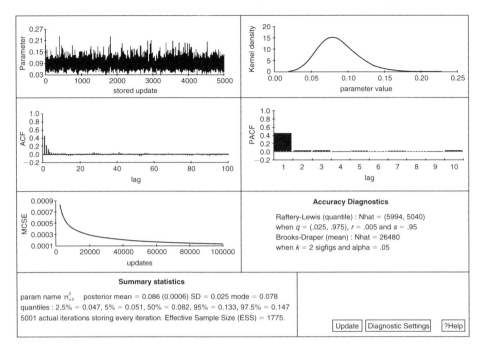

Figure 2.15 MCMC diagnostics from MLwiN.

We also see, from the top left-hand box, that the chain appears to have mixed well, and the box below this shows a first order correlation of just over 0.4 with little evidence of a second order partial correlation in the middle box in the right hand column. This gives rise to an effective sample size of about 1800.

We now look at the variance partition coefficient. The value of this derived from the final estimates of the variances, is 0.159 for the ML estimates, 0.179 for the uniform priors and 0.168 for the gamma priors. If we calculate the variance partition coefficient for every iteration we find that, for the gamma priors, the mean value is 0.167 with a standard deviation of 0.042 and a 95% interval of (0.097, 0.263). Any function of the parameters can be 'monitored' in this way, including the residuals. Thus, for each level 2 residual we will have a chain and this allows us to place interval estimates about these. This will allow us, for example, to present a 'caterpillar graph' as in Figure 2.14. If we wish to provide simultaneous intervals for pairwise comparisons of schools, then we can apply a scaling factor to the MCMC intervals, as discussed in section 2.12.4 based upon a normal assumption; in practice a value of 1.4 appears to work well in many cases and a more detailed computation may be unnecessary.

In some cases we may be more interested in presenting the rankings of schools rather than their means, for example if the results are to be interpreted for the purpose of identifying which order they come in. In this case we would form a ranking of the residuals at each iteration and evaluate the average ranking of each school, together with any interval estimate, over the chain. These final average ranks will not necessarily order the schools the same as the means, but the correlation between them will normally be high. There is a useful discussion of whether to use rankings or means in the discussion section of the paper by Goldstein and Spiegelhalter (1996). One advantage

Table 2.11 MCMC estimates for model in Table 2.8 with Normal and *t*-distributed level 1 residuals (standard errors are in parentheses [and 95% interval for degrees of freedom]). Gamma priors

	t-*distribution*	*Normal distribution*
Fixed		
Intercept	0.116	0.130
8-year score	0.679 (0.026)	0.669 (0.027)
Gender (boys–girls)	−0.044 (0.049)	−0.047 (0.050)
Social class (non-manual–manual)	−0.134 (0.058)	−0.138 (0.053)
Random		
Level 2: σ_{u0}^2	0.086 (0.025)	0.086 (0.026)
Level 1: σ_{e0}^2	0.354 (0.036)	0.426 (0.023)
Effective number of parameters	38.9	38.1
Degrees of freedom	16.2 (12.5) (5.9, 45.5)	
DIC	1472.6	1481.1

of presenting interval estimates for rankings is that it allows us to say immediately what percentage of the overall distribution for schools is covered. For example, the school with the smallest residual rank has a set of rankings for the chain of 5000 iterations extending from the lowest rank of 1 to its highest of 20 but the symmetrical 95% interval is from rank 1 to 9 only. By contrast, for the school with the highest average rank the interval is from 42 to 48 . The correlation between the residual ranks and the means is 0.99.

Finally, we look at allowing for non-Normality of the level 1 residuals. We shall allow them to have a general *t*-distribution since this is symmetrical and allows any positive or negative value. The Normal distribution is the limiting case where the degrees of freedom parameter goes to infinity. Recall that we have forced the response variable to have a Normal distribution by use of Normal equivalent scores so we can regard the fitting of the *t*-distribution as a way of checking whether the Normality assumption carries through to the residuals. The estimation is done by first running a Gibbs estimation with gamma priors in MLwiN (Rasbash et al., 2000) and then using that program's facility for producing input to the general purpose MCMC program BUGS (Spiegelhalter et al., 2000) to fit the *t*-distribution using MH sampling. The degrees of freedom parameter for the *t*-distribution was also estimated, with a flat prior distribution (uniform in the range (1,10 000)). The number of iterations was set to 5000. The results are given in Table 2.11.

The degrees of freedom parameter estimate has a mean of 16.2, suggesting a fairly 'heavy-tailed' distribution, although with a wide interval estimate. Note that the estimated density for this parameter, as shown in Figure 2.16, is very skew so that the standard error should not be used to provide an interval estimate. The fixed parameter estimates do change slightly when we fit the *t*-distribution and we see that the DIC shows a modest improvement.

We shall be applying MCMC methods in subsequent chapters alongside IGLS estimation and for certain models we shall see that MCMC has particular advantages. For purposes of data-driven model selection, MCMC methods can be very time consuming. Generally therefore, we can use likelihood-based methods to choose one or two suitable models that fit the data and then carry out the MCMC estimation. Likewise, the model diagnostic procedures discussed in sections 2.8 and 2.9 will often be time consuming and not really feasible for use with MCMC.

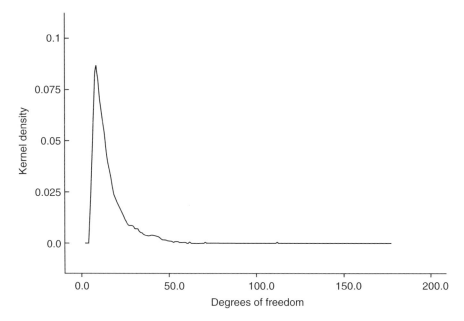

Figure 2.16   Estimated kernel density for degrees of freedom parameter in Table 2.11.

2.14   Data augmentation

Finally we look briefly at a technique that can be viewed as a special kind of Gibbs sampler (see Appendix 2.4). Data augmentation (Tanner and Wong, 1987) aims to provide an estimate of the full posterior distribution of the parameters in the same kind of way. The algorithm we shall consider essentially consists of two steps, the imputation (I) step and the posterior (P) step and we shall use the term IP algorithm. It is motivated, like the EM algorithm, by having 'missing data'; in multilevel models these typically are the unknown random effects but, for example, randomly missing responses can be included also. Schafer (1997) gives examples.

The IP algorithm has two steps:

- **Step 1**
  Draw a sample of the random effects given the data and current values of the model parameters. This is essentially step 2 of the Gibbs algorithm (see Appendix 2.4).
- **Step 2**
  Draw a sample from the conditional distribution of the parameters given the data and current parameter values and the random effects.

In effect this algorithm can be regarded as a form of Gibbs sampling where all the parameters are sampled in a single step, although this step can also be broken down into smaller steps fitting random and fixed parameters in sequence. It requires a prior distribution specification, which can be diffuse, and is formally a Bayesian procedure.

An interesting application to the analysis of large cross-classified data (see Chapter 11) is given by Clayton and Rasbash (1999). They consider a set of two parallel IP algorithms or 'wings', one for each cross-classification (A and B) with the following steps:

- **Step 1 for classification A**
  Given the current set of estimated residuals from classification B fit an ordinary multilevel model using, for example, IGLS.
- **Step 2 for classification A**
  Sample the parameters from their new conditional posterior distribution. This is essentially the combined steps 1, 3, 4 of our Gibbs sampler.
- **Step 3 for classification A**
  Sample a set of residuals from their new conditional posterior distribution. This is step 2 of our Gibbs sampler.

These three steps are repeated for classification B and the process run, with a suitable burn in, to produce a chain of parameters as in the standard Gibbs sampler. Clayton and Rasbash note that this method is especially suited to large datasets where the full IGLS and RIGLS methods become computationally inefficient. They also point out that in such cases, it may be unnecessary to sample the random parameters, so speeding up the computations, which is particularly important in large data problems.

Step 1 is required to set up the conditions for sampling the parameters from each wing, but a standard Gibbs algorithm can be implemented by sampling residuals and parameters from all the higher level units simultaneously, although this will generally be much slower.

Appendix 2.1

The general structure and maximum likelihood estimation for a multilevel model

We illustrate the general structure using a 2-level model. We have

$$
\begin{aligned}
Y &= XB + E \\
Y &= \{y_{ij}\}, \quad X = \{X_{ij}\}, \quad X_{ij} = \{x_{0ij}, x_{1ij}, \dots, x_{pij}\} \\
E &= E_1 + E_2 = \{e_{ij}\}, \quad e_{ij} = e_{ij}^{(1)} + e_j^{(2)}, \\
e_{ij}^{(1)} &= \sum_{h=0}^{q_1} z_{hij}^{(1)} e_{hij}^{(1)}, \quad e_j^{(2)} = \sum_{h=0}^{q_2} z_{hij}^{(2)} e_{hj}^{(2)}
\end{aligned}
\tag{2.1.1}
$$

We will also write simply

$$
e_{ij}^{(1)} = e_{ij}, \quad e_j^{(2)} = u_j
$$

$$
Y = X\beta + Z^{(2)}u + Z^{(1)}e
$$

or

$$
Y = X\beta + Z_u u + Z_e e
$$

The residual matrices E_1, E_2 have expectation zero with

$$
\begin{aligned}
E(E_1 E_1^T) &= V_{2(1)}, \quad E(E_2 E_2^T) = V_{2(2)} \\
E(E_1 E_2^T) &= 0, \quad V_2 = V_{2(1)} + V_{2(2)}
\end{aligned}
\tag{2.1.2}
$$

In the standard model the level 1 residuals are assumed independent across level 1 units, so that $V_{2(1)}$ is diagonal with ij-th element

$$
\mathrm{var}(e_{ij}) = \sigma_{eij}^2 = z_{ij}^{(1)^T} \Omega_e z_{ij}^{(1)}, \quad \Omega_e = \mathrm{cov}(e_h^{(1)})
$$

The level 2 residuals are assumed independent across level 2 units and $V_{2(2)}$ is block-diagonal with j-th block

$$
V_{2(2)j} = z_j^{(2)^T} \Omega_u z_j^{(2)}, \quad \Omega_u = \mathrm{cov}(e_h^{(2)})
$$

The j-th block of V_2 is therefore given by

$$
V_{2j} = \oplus_i \sigma_{eij}^2 + V_{2(2)j}
\tag{2.1.3}
$$

where $\oplus$ is the direct sum operator.

For some of the models dealt with in later chapters, such as the time series models of Chapter 6, the requirement of independence among the residuals for the level 1 units is relaxed. In this case the first term on the right hand side of (2.1.3) is replaced by the particular structure of $V_{2(1)}$.

For known V_2 and omitting the subscript for convenience, the generalized least squares estimate of the fixed coefficients is

$$\hat{\beta} = (X^T V^{-1} X)^{-1} X^T V^{-1} Y \tag{2.1.4}$$

with covariance matrix

$$(X^T V^{-1} X)^{-1}$$

For known β we form

$$Y^* = \tilde{Y}\tilde{Y}^T, \quad \tilde{Y} = Y - X\beta = E_1 + E_2 \tag{2.1.5}$$

and we have $E(Y^*) = V$. We now write

$$Y^{**} = vec(Y^*)$$

where vec is the vector operator stacking the columns of Y^* underneath each other. We can now write a linear model involving the random parameters, that is the elements of Ω_u, Ω_e, as follows

$$E(Y^{**}) = Z^* \theta \tag{2.1.6}$$

where Z^* is the design matrix for the random parameters. An example of such a design matrix for a simple variance components model is given in Chapter 2. We now carry out a generalized least squares analysis to estimate θ, namely

$$\hat{\theta} = (Z^{*^T} V^{*^{-1}} Z^*)^{-1} Z^{*^T} V^{*^{-1}} Y^{**}, \quad V^* = V \otimes V \tag{2.1.7}$$

where $\otimes$ is the Kronecker product. The covariance matrix of $\hat{\theta}$ is given by

$$(Z^{*^T} V^{*^{-1}} Z^*)^{-1} Z^{*^T} V^{*^{-1}} \text{cov}(Y^{**}) V^{*^{-1}} Z^* (Z^{*^T} V^{*^{-1}} Z^*)^{-1}$$

Now we have

$$Y^{**} = vec(\tilde{Y}\tilde{Y}^T) = \tilde{Y} \otimes \tilde{Y}$$

Using a standard result (for example section 12.3 in Searle et al. 1992) we have

$$\text{cov}(\tilde{Y} \otimes \tilde{Y}) = (V \otimes V)(I + S_N)$$

where $V \otimes V = V^*$ and S_N is the vec permutation matrix.

As Goldstein and Rasbash (1992) note, the matrix A where $Z^* = vec(A)$ is symmetrical, and hence

$$V^{*^{-1}} Z^* = (V^{-1} \otimes V^{-1}) vec(A) = vec(V^{-1} A V^{-1})$$

and $V^{-1} A V^{-1}$ is symmetric so that, using a standard result, we have

$$S_N V^{*^{-1}} Z^* = V^{*^{-1}} Z^*$$

and after substituting in the above expression for $\text{cov}(\hat{\theta})$ we obtain

$$\text{cov}(\hat{\theta}) = 2(Z^{*^T} V^{*^{-1}} Z^*)^{-1} \qquad (2.1.8)$$

The iterative generalized least squares (IGLS) procedure (Goldstein, 1986) iterates between (2.1.4) and (2.1.7) using the current estimates of the fixed and random parameters. Typical starting values for the fixed parameters are those from an ordinary least squares analysis. At convergence, assuming multivariate Normality, the estimates are maximum likelihood.

The IGLS procedure produces biased estimates in general and this can be important in small samples. Goldstein (1989a) shows how a simple modification leads to restricted iterative generalized least squares (RIGLS) or restricted maximum likelihood (REML) estimates which are unbiased. If we rewrite (2.1.5) using the *estimates* of the fixed parameters $\hat{\beta}$ we obtain

$$E(Y^*) = V_2 - X\text{cov}(\hat{\beta})X^T = V_2 - X(X^T V_2^{-1} X)^{-1} X^T \qquad (2.1.9)$$

By taking account of the sampling variation of the $\hat{\beta}$ we can obtain an unbiased estimate of V_2 by adding the second term in (2.1.9), the 'hat' matrix, from Y^* at each iteration until convergence. In the case where we are estimating a variance from a simple random sample this becomes the standard procedure for using the divisor $n - 1$ rather than n to produce an unbiased estimate.

Full details of efficient computational procedures for carrying out all these calculations are given by Goldstein and Rasbash (1992).

Appendix 2.2

Multilevel residuals estimation

2.2.1 Shrunken estimates

Denote the set of m_h distinct residuals at level h ($h > 1$) in a multilevel model by

$$u^{(h)} = \{u_1^{(h)}, \ldots, u_{m_h}^{(h)}\}, \quad u_i^{(h)^T} = \{u_{i1}^{(h)}, \ldots, u_{in_h}^{(h)}\} \qquad (2.2.1)$$

where n_h is the number of level h units. Since the residuals at any level are independent of those at any other level for each residual vector we require the posterior or predicted residual estimates given by

$$\hat{u}_i^{(h)} = E(u_i^{(h)}|\tilde{Y}, V)$$

where $\tilde{Y} = Y - X\beta$. We consider the regression of the set of all residuals $u^{(h)}$ on $\tilde{Y}$ which gives the estimator

$$\hat{u}^{(h)} = R_h^T V^{-1} \tilde{Y} \qquad (2.2.2)$$

where R_h is block-diagonal, each block corresponding to a level h unit and for the j-th block given by

$$Z_j^{(h)'} \Omega_h$$

where $Z_j^{(h)}$ is the matrix of explanatory variables for the random coefficients at level h. We obtain consistent estimators by substituting sample estimates of the parameters in (2.2.2). These estimates are linear functions of the responses and their unconditional covariance matrix is given by

$$R_h^T V^{-1} (V - X(X^T V^{-1} X)^{-1} X^T) V^{-1} R_h \qquad (2.2.3)$$

The second term in (2.2.3) derives from considering the sampling variation of the estimates of the fixed coefficients and can be ignored in large samples and we obtain a consistent estimator by substituting parameter estimates in

$$R_h^T V^{-1} R_h$$

Note that there are no covariances across units. Where we wish to study the distributional properties of standardized residuals for diagnostic purposes then the unconditional covariance matrix (2.2.3) should be used to standardize the estimated residuals. If, however, we wish to make inferences about the true $u_i^{(h)}$, for example to construct confidence intervals or test differences, then we require the conditional or 'comparative' covariance matrix of $\hat{u}^{(h)}|u^{(h)}$ or $E[(\hat{u}^{(h)} - u^{(h)})(\hat{u}^{(h)} - u^{(h)})^T]$, which is given by substituting parameter estimates in

$$S_h - R_h^T V^{-1}(V - X(X^T V^{-1}X)^{-1}X^T)V^{-1}R_h \tag{2.2.4}$$

where S_h is the block-diagonal matrix where each block corresponds to a level h unit is Ω_h. We note that no account is taken of the sampling variability associated with the estimates of the random parameters in (2.2.3) or (2.2.4). Thus with small numbers of units, a procedure such as bootstrapping should be used to estimate these covariance matrices (see Chapter 3) or a delta method approximation can be used as shown below.

If we write an $(l + 1)$ level model in the form, using an extension of the notation in Appendix 2.1, as follows

$$\begin{aligned} Y &= X\beta + ZU + Z^{(1)}e, \\ Z &= \{Z^{(l)}, Z^{(l-1)}, \ldots, Z^{(2)}\}, \quad U^T = \{u^{(l)}, u^{(l-1)}, \ldots u^{(2)}\} \end{aligned} \tag{2.2.5}$$

then we can derive the following form for the residual estimator that is computationally convenient. For the j-th level h residual we have

$$\hat{u}_j^{(h)} = \{\Omega^{(h)-1} + z_j^{(h)^T} V_{1j}^{-1} z_j^{(h)}\}^{-1} z_j^{(h)^T} V_{1j}^{-1} \tilde{y}_j^{(h)} \tag{2.2.6}$$

with a consistent estimator for the comparative covariance matrix as

$$\{\Omega^{(h)-1} + z_j^{(h)^T} V_{1j}^{-1} z_j^{(h)}\}^{-1}$$

where the subscript j refers to the set of level 1 units in the j-th level h unit. Corresponding to (2.2.3) we have the following additional correction factor, to be subtracted from this covariance matrix, that takes into account the sampling variation of the fixed coefficients

$$\begin{aligned} &\{\Omega^{(h)-1} + z_j^{(h)^T} V_{1j}^{-1} z_j^{(h)}\}^{-1} z_j^{(h)^T} V_{1j}^{-1} \left[\sum_h z_j^{(h)} \Omega^{(h)} z_j^{(h)^T} \right] V_{1j}^{-1} z_j^{(h)} \\ &\times \{\Omega^{(h)-1} + z_j^{(h)^T} V_{1j}^{-1} z_j^{(h)}\}^{-1} \end{aligned} \tag{2.2.7}$$

2.2.2 Delta method estimators for the covariance matrix of residuals

Consider the case of a 2-level Normal model

$$y = X\beta + Z_u u + Z_e e$$

where the estimates are given by (2.2.2) and variance estimates by (2.2.4).
Using a well known equality we have

$$\text{var}(\hat{u}|y, \beta, \theta) = E[\text{var}(\hat{u}|y, \beta, \theta)] + \text{var}[E(\hat{u}|y, \beta, \theta)] \tag{2.2.8}$$

where the terms on the right-hand side of (2.2.8) are regarded as functions of the model parameters and evaluated at the sample estimates. The comparative covariance matrix given by (2.2.4) is the first term in (2.2.8). For the second term we shall use the first order approximation derived from the Taylor expansion about $E(\hat{\theta}) = \theta$, for the covariance matrix of a function, namely

$$\text{cov}[g(\hat{\theta})] \approx \left(\frac{\partial g}{\partial \theta}\right)^T \text{cov}(\hat{\theta}) \left(\frac{\partial g}{\partial \theta}\right)\bigg|_{\hat{\theta}} \qquad (2.2.9)$$

In some circumstances we may wish to have a better approximation, in which case, assuming multivariate Normality, we obtain the additional contribution, evaluated at the sample estimates

$$\frac{1}{4} \left(\frac{\partial^2 g}{\partial \theta^2}\right)^T [2A_1 + A_2] \left(\frac{\partial^2 g}{\partial \theta^2}\right)\bigg|_{\hat{\theta}}$$

$$A_1 = \{a_{ij}^2\} \quad \text{where } \text{cov}(\hat{\theta}) = \{a_{ij}\}$$

$$A_2 = aa^T \quad \text{where } a = \{a_{ii}\}$$

For $\hat{u}$ as a function of the random parameters θ, we have

$$d_k^T = \frac{\partial g}{\partial \theta_k} = \left[-\Omega_u Z^T V^{-1} \frac{\partial V}{\partial \theta_k} V^{-1} + \frac{\partial \Omega_u}{\partial \theta_k} Z^T V^{-1}\right] \tilde{y}$$

$$\frac{\partial \Omega_u}{\partial \theta_k} = 0 \quad \text{if } \theta_k \text{ not at level 2.} \qquad (2.2.10)$$

Note that the elements of $\partial V/\partial \theta_k$ are just the elements of the design vector for the parameter θ_k and that

$$\frac{\partial V^{-1}}{\partial \theta_k} = -V^{-1} \frac{\partial V}{\partial \theta_k} V^{-1}$$

The row vector d_k has q elements, one for each residual at level 2 with $d = \{d_k\}$ a $t \times r_u$ matrix where t is the total number of random parameters. The second adjustment term in (2.2.8) is therefore

$$d^T \text{cov}(\hat{\theta}) d$$

This procedure for the variance of the estimated residuals is essentially equivalent to that proposed by Kass and Steffey (1989) who give an alternative derivation using the Laplace method. These authors also consider the extra adjustment term based upon the next term in the Taylor expansion as above.

Appendix 2.3

The EM algorithm

To illustrate the procedure, consider the 2-level variance components model

$$y_{ij} = (X\beta)_{ij} + u_j + e_{ij}, \quad \text{var}(e_{ij}) = \sigma_e^2, \quad \text{var}(u_j) = \sigma_u^2 \tag{2.3.1}$$

The vector of level 2 residuals is treated as missing data and the 'complete' data therefore consists of the observed vector Y and the u_j treated as responses. The joint distribution of these, assuming Normality, and using our standard notation is

$$\begin{bmatrix} Y \\ u \end{bmatrix} = N \left\{ \begin{bmatrix} X\beta \\ 0 \end{bmatrix}, \begin{bmatrix} V & J^T \sigma_u^2 \\ \sigma_u^2 J & \sigma_u^2 I \end{bmatrix} \right\} \tag{2.3.2}$$

This generalizes readily to the case where there are several random coefficients. If we denote these by β_j we note that some of them may have zero variances. We can now derive the distribution of $\beta_j | Y$ in Appendix 2.2, and we can also write down the Normal log-likelihood function for (2.3.2) with a general set of random coefficients, namely

$$\log(L) \propto -N \log(\sigma_e^2) - m \log|\Omega_u| - \sigma_e^{-2} \sum_{ij} e_{ij}^2 - \sum_j \beta_j^T \Omega_u^{-1} \beta_j$$

$$\Omega_u = \text{cov}(\beta_j) \tag{2.3.3}$$

Maximizing this for the random parameters we obtain

$$\hat{\sigma}_e^2 = N^{-1} \sum_{ij} e_{ij}^2$$

$$\hat{\Omega}_u = m^{-1} \sum_j \beta_j \beta_j^T \tag{2.3.4}$$

where m is the number of level 2 units. We do not know the values of the individual random variables. We require the expected values, conditional on the Y and the current parameters, of the terms under the summation signs, these being the sufficient statistics. We then substitute these expected values in (2.3.4) for the updated random parameters. These conditional values are based upon the 'shrunken' predicted values and their (conditional) covariance matrix, given in Appendix 2.2. With these updated values of the random parameters we can form V and hence obtain the updated estimates for the fixed parameters using generalized least squares. We note that the expected values of

the sufficient statistics can be obtained using the general result for a random parameter
vector θ

$$E(\theta\theta^T) = \text{cov}(\theta) + [E(\theta)][E(\theta)]^T \tag{2.3.5}$$

The prediction is known as the E (expectation) step of the algorithm and the computations in (2.3.4) the M (maximization) step. Given starting values, based upon OLS,
these computations are iterated until convergence is obtained. Convenient computational formulae for computing these quantities at each iteration can be found in Bryk
and Raudenbush (2002). Meng and Dyke (1998) describe fast implementations of the
EM algorithm.

Appendix 2.4

MCMC sampling

MCMC estimation proceeds, at each iteration, by considering each component in turn and generating a random sample from the distribution of that component assuming the current values of the remaining components (known as the conditional posterior distribution). Under general conditions, after the 'burn in', the resulting chain of parameter vectors can be considered as a (serially correlated) random sample from the joint posterior distribution of the parameters. To illustrate the computations we consider the following variance components model:

$$y_{ij} = (X\beta)_{ij} u_j + e_{ij}, \quad \text{var}(e_{ij}) = \sigma_e^2, \quad \text{var}(u_j) = \sigma_u^2 \qquad (2.4.1)$$

2.4.1 Gibbs sampling

The Gibbs algorithm proceeds through a series of steps as follows:

Step 1

Sample a new set of fixed effects from the conditional posterior distribution

$$p(\beta|y, \sigma_u^2, \sigma_e^2, u) \propto L(y; \beta, u, \sigma_e^2) p(\beta)$$

A suitable diffuse prior is $p(\beta) \propto 1$

$$p(\beta|y, \sigma_u^2, \sigma_e^2, u) \propto \left(\frac{1}{\sigma_e^2}\right)^{N/2} \prod_{i,j} \exp\left[-\frac{1}{2\sigma_e^2}(y_{ij} - u_j - (X\beta)_{ij})^2\right]$$

so that we sample from

$$\beta \sim MVN(\hat{\beta}, \hat{D})$$

$$\hat{\beta} = \left[\sum_{i,j} X_{ij}^T X_{ij}\right]^{-1} \left[\sum_{i,j} X_{ij}^T (y_{ij} - u_j)\right]$$

$$\hat{D} = \sigma_e^2 \left[\sum_{i,j} X_{ij}^T X_{ij}\right]^{-1}$$

if $p(\beta) \sim MVN(0, V)$

$$\beta \sim MVN(\hat{\beta}^*, \hat{D}^*)$$

$$\hat{\beta}^* = \left[\sum_{i,j} X_{ij}^T X_{ij} + \sigma_e^2 V^{-1} \right]^{-1} \left[\sum_{i,j} X_{ij}^T (y_{ij} - u_j) \right]$$

$$\hat{D}^* = \sigma_e^2 \left[\sum_{i,j} X_{ij}^T X_{ij} + \sigma_e^2 V^{-1} \right]^{-1}$$

In the above and subsequent equations the parameter terms on the right-hand sides represent the current values and those on the left-hand sides the new updated ones; for simplicity we omit any indexing of the iterations where this leads to no ambiguity. The uniform prior distribution, extending over the whole real line, is 'improper' in the sense that it is not a true probability distribution. Nevertheless, for independent Normal priors the variance matrix, V, for the fixed parameters is diagonal and if the elements are very large the posterior distribution for β is equivalent to assuming uniform prior distributions. The important requirement is that the posterior distribution exists.

Note that we could sample each coefficient in turn but this would tend to produce more highly correlated chains compared to sampling them as a block where terms such as $\left[\sum_{i,j} X_{ij}^T X_{ij} \right]$ do not change and can be stored for use in each iteration.

Step 2

Sample a new set of residuals

Each residual u_j is assumed to have a prior distribution $u_j \sim N\left(0, \sigma_u^2\right)$ which leads to the following posterior distribution, where n_j is the number of level 1 units in the j-th level 2 unit

$$p(u_j | y, \sigma_u^2, \sigma_e^2) \propto \left(\frac{1}{\sigma_e^2} \right)^{n_j/2} \prod_{i=1}^{n_j} \exp\left[-\frac{1}{2\sigma_e^2} (y_{ij} - (X\beta)_{ij} - u_j)^2 \right]$$

so that we now sample from

$$u_j \sim N(\hat{u}_j, \hat{D})$$

$$\hat{u}_j = [n_j + \sigma_e^2 \sigma_u^{-2}]^{-1} \left[\sum_{i=1}^{n_j} (y_{ij} - (X\beta)_{ij}) \right]$$

$$\hat{D} = \sigma_e^2 [n_j + \sigma_e^2 \sigma_u^{-2}]^{-1}$$

Where there are $p \, (>1)$ random coefficients with explanatory variable matrix Z, this step is modified by sampling the residuals from a p-variate normal distribution with

$$\hat{u}_j = \left[\sum_{i=1}^{n_j} Z_{ij}^T Z_{ij} + \sigma_e^2 \Omega_u^{-1} \right]^{-1} \left[\sum_{i=1}^{n_j} Z_{ij}^T (y_{ij} - (X\beta)_{ij}) \right]$$

$$\hat{D} = \sigma_e^2 \left[\sum_{i=1}^{n_j} Z_{ij}^T Z_{ij} + \sigma_e^2 \Omega_u^{-1} \right]^{-1}$$

These expressions are equivalent to those given in Appendix 2.2.

Step 3

Sample a new level 2 variance

An issue arises here over the choice of a suitable diffuse prior distribution. A detailed discussion of this choice is given by Browne (1998) and we will here just mention two possibilities, a uniform prior as in the case of the fixed parameters and an inverse gamma prior. The former choice will tend often to produce positively biased estimates and the latter is commonly used. Rather than sampling the variance directly it is simpler to sample its inverse σ_u^{-2}, referred to as the 'precision' which is assumed to have the gamma prior distribution $p(\sigma_u^{-2}) \sim gamma(\varepsilon, \varepsilon)$. We sample from

$$\sigma_u^{-2} \sim gamma(a_u, b_u)$$

$$a_u = (J + 2\varepsilon)/2 \quad b_u = \left(2\varepsilon + \sum_{j=1}^{J} u_j^2 \right)$$

where J is number of level 2 units.

For a uniform prior for σ_u^2 we sample from a gamma distribution with

$$a_u = (J - 2)/2, \quad b_u = \sum_{j=1}^{J} u_j^2$$

Where there are $p \, (>1)$ random coefficients we need to sample a new covariance matrix Ω_u and corresponding to the two choices for a single variance are a set of uniform priors or an inverse Wishart prior. We now sample from

$$\Omega_u^{-1} \sim Wishart(v_u, S_u)$$

$$v_u = J + v_p, \; S_u = \left(\sum_{j=1}^{J} u_j^T u_j + S_p^{-1} \right)^{-1}$$

where u_j is the row vector of residuals for the j-th level 2 unit and the prior $p(\Omega_u^{-1}) \sim Wishart(v_p, S_p)$.

A 'minimally informative' or 'maximally diffuse' choice for the prior would be to take v_p equal to the order of Ω_u and S_p equal to a value chosen to be close to the final estimate multiplied by v_p; typically we can use the maximum likelihood or approximate maximum likelihood estimate for S_p that is also used for the MCMC starting values. Choosing $v_p = -3, S_p = 0$ is equivalent to choosing a uniform prior for Ω_u.

Step 4

Sample a new level 1 variance
   This is similar to the procedure for a single level 2 variance. We sample from

$$\sigma_e^{-2} \sim gamma(a_e, b_e)$$

$$a_e = (N + 2\varepsilon)/2 \quad b_e = \left(2\varepsilon + \sum_{i,j} e_{ij}^2\right)$$

For a uniform prior we can sample from a gamma distribution with

$$a_e = (N - 2)/2, \quad b_e = \sum_{i,j} e_{ij}^2$$

Step 5

Compute the level 1 residuals
   We calculate the e_{ij} by subtraction using (2.4.1).

2.4.2   Metropolis Hastings (MH) sampling

Consider step 1 in (2.4.1) for the fixed coefficients. Given the current values of the fixed coefficients we form a proposal distribution and sample a new set of coefficients from this distribution. We then apply a rule for accepting this new set or rejecting it and remaining with the current values. Ideally we would wish to have a proposal distribution that minimizes the autocorrelation and hence the number of iterations for a given accuracy. We shall consider how this may be achieved after describing the basic principle.

   One simple approach is to sample from a multivariate normal distribution of the following form:

$$\beta_{(t)} \sim MVN(\beta_{(t-1)}, D) \tag{2.4.2}$$

where D, for example, is the covariance matrix of the coefficient estimates, but see below for a modification. Having obtained a new set, say β^*, we need an acceptance rule and this is based upon the following ratio:

$$r_t = p(\beta^*, \sigma_u^2, \sigma_e^2, u|y)/p(\beta_{(t-1)}, \sigma_u^2, \sigma_e^2, u|y) \tag{2.4.3}$$

where these probability densities have the same form as the conditional probabilities in step 1. This procedure, which conditions the proposal distribution on the current parameter values, is known as a *random walk Metropolis sampler*.

   We note in passing that the multivariate normal distribution is symmetric in $\beta_{(t)}, \beta_{(t-1)}$ that is, $p(\beta_{(t)} = a|\beta_{(t-1)} = b) = p(\beta_{(t)} = b|\beta_{(t-1)} = a)$. Where asymmetric distributions are used the above ratio has to be further multiplied by the so called *Hastings ratio* $p(\beta_{(t-1)}|\beta^*)/p(\beta^*|\beta_{(t-1)})$.

We define the acceptance probability for the new set as $\alpha_t = min(1, r_t)$. If this is 1 then the new set is accepted into the chain. If it is less than 1 then it is accepted with probability α_t using a suitable random mechanism, for example generating a uniform random number in (0,1) and accepting if this is less than α_t. If the new set is not accepted then the current $(t-1)$ set is used. Thus, unlike Gibbs sampling, it is possible for the chain to stay with the same set of values for several iterations.

An efficient MH sampler attempts to avoid this as far as possible, even though the MH sampler in principle will produce satisfactory estimates if allowed to run for long enough; the key is to find a suitable proposal distribution. Gelman et al. (1996) suggest using $(5.66/d)^*D$, where d is the dimension of the covariance matrix D. A somewhat more satisfactory procedure (see Browne and Draper, 2000) is to specify a desired acceptance rate (e.g. 50%) and adjust the proposal distribution scaling factors for each of the parameters. This can be done during a preliminary adaptive stage of the MCMC chain estimation before the 'burn in'.

For the variance matrix a standard proposal is to sample the inverse of the matrix from a Wishart distribution and for a single variance to sample the inverse (precision) from a gamma distribution. We can again use an adaptive procedure to determine a scaling 'degrees of freedom' parameter. Browne et al. (2001c) discuss various possibilities with examples, including the special case where the level 1 variance is a linear function of covariates (see Chapter 3 for a discussion of complex level 1 variation).

Generally, the variance and covariance sampling can be carried out using Gibbs and this gives an efficient hybrid algorithm (see Browne and Draper, 2000) which is a mixture of MH and Gibbs sampling steps.

2.4.3 Hierarchical centring

Consider the three-level model

$$y_{ijk} = (X\beta)_{ijk} + v_k + u_{jk} + e_{ijk}$$
$$v_k \sim N(0, \sigma_v^2), \quad u_{jk} \sim N(0, \sigma_u^2), \quad e_{ijk} \sim N(0, \sigma_e^2) \qquad (2.4.4)$$

which we can write as

$$y_{ijk} = \beta_0 + (X'\beta')_{ijk} + u_{jk}^* + e_{ijk}$$
$$u_{jk}^* \sim N(v_k, \sigma_u^2), \quad v_k \sim N(\beta_0, \sigma_v^2), \quad e_{ijk} \sim N(0, \sigma_e^2) \qquad (2.4.5)$$

where, instead of sampling u_{jk}, v_k separately, we centre v_k on β_0 and sample the u_{jk}^* conditionally on the v_k. This procedure, with extensions to more complex models, will often greatly improve convergence (Gelfand et al., 1995; Browne, 2002).

3

Three-Level Models and More Complex Hierarchical Structures

3.1 Complex variance structures

In all the models of Chapter 2 we have assumed that a single variance describes the random variation at level 1. At level 2 we have introduced a more complex variance structure, by allowing regression coefficients to vary across level 2 units. The modelling and interpretation of this complex variation was in terms of randomly varying coefficients. An alternative way of looking at this structure, however, is that the level 2 variance is a function of the explanatory variables that have a random coefficient. Thus, for the model of Table 2.1, if we add a random coefficient for the 8-year score to give the model

$$y_{ij} = \beta_0 + \beta_1 x_{1ij} + (u_{0j} + u_{1j}x_{1ij} + e_{0ij})$$
$$\text{var}(e_{0ij}) = \sigma_{e0}^2 \tag{3.1}$$

the level 2 variance is given by

$$\text{var}(u_0 + u_1 x_{1ij}) = \sigma_{u0}^2 + 2\sigma_{u01}x_{1ij} + \sigma_{u1}^2 x_{1ij}^2 \tag{3.2}$$

This is a quadratic function of x_1 and if we had further random coefficients the variance would be a function of all of the associated explanatory variables. The function is plotted in Figure 3.1. Although it does not occur here, the level 2 variance function can sometimes become negative within the range of the data. This is due to sampling variation since there is no restriction in the standard IGLS estimation that forces the variance function to be non-negative. One way of avoiding this problem is to have a nonlinear function such as $e^{-f(x)}$ where $f(x)$ is a polynomial such as in (3.2) and we shall deal with this case in Chapter 8.

Now we look at how we can model the variation at level 1 as a function of explanatory variables and how this can give substantively interesting interpretations. In Figure 2.1 we saw that the level 1 residual variation appeared to decrease with increasing 8-year maths score. Since we shall now consider several random variables at each level the notation used in Chapter 2 needs to be extended. Extending the notation in (2.6) for the total level 2 random effect we write

$$u_j = \sum_{h=0}^{l_2} u_{hj}z_{hij} \tag{3.3}$$

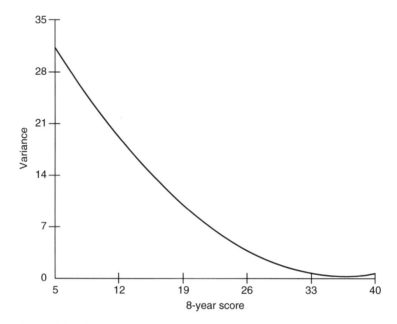

Figure 3.1 Level 2 variance by 8-year score.

where there are l_2 random effects each with an explanatory variable (z_{hij}) where, by convention, z_{0ij} refers to the constant ($=1$) defining a basic or intercept variance term. We can also write a similar expression for the level 1 random effect as

$$e_{ij} = \sum_{h=0}^{l_1} e_{hij} z_{hij} \qquad (3.4)$$

In both cases the z_{hij} may be either level 1 or level 2 defined explanatory variables. In the standard case we considered in Chapter 2 there is a single variable $z_{0ij} = 1$ and an associated level 1 variance σ_{e0}^2 as in (3.1).

Suppose now that we introduce an extra random term, $e_{1ij}x_{1ij}$ into (3.1) to give

$$y_{ij} = \beta_0 + \beta_1 x_{1ij} + (u_{0j} + u_{1j}x_{1ij} + e_{0ij} + e_{1ij}x_{1ij})$$
$$\text{var}(e_{ij}) = \text{var}(e_{0ij} + e_{1ij}x_{1ij}) = \sigma_{e0}^2 + 2\sigma_{e01}x_{1ij} + \sigma_{e1}^2 x_{1ij}^2 \qquad (3.5)$$

so that we have made the level 1 variance a quadratic function of x_1. Table 3.1 shows the results of fitting this model and Figure 3.2 shows the level 1 variance function for model A.

This is just the form of relationship we would expect given the 'ceiling' effect for the 11-year score. While this specification of the level 1 variance as a polynomial function is analogous to the specification for the level 2 variance, we do not have an alternative interpretation in terms of random slopes as we did at level 2. In other words, the parameter σ_{e1}^2 cannot be interpreted as the variance of a coefficient; it is simply the coefficient of the quadratic term describing the level 1 *complex variance function*.

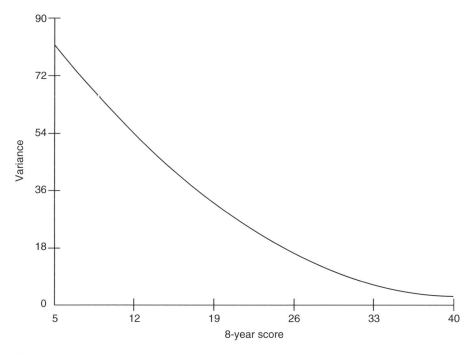

Figure 3.2 Level 1 variance by 8-year score.

Note that, although we have retained the same notation for the parameters of the variance function, they no longer are to be regarded as variances or covariances of measured or unmeasured variables. Of course, where a coefficient is made random at a level higher than that at which the explanatory variable itself is defined, then the resulting variance (and covariance) *can* be interpreted as the between-higher-level unit variance of the within-unit relationship described by the coefficient.

We can consider other functions, for example a linear function of x_1. Thus, if we write

$$y_{ij} = \beta_0 + \beta_1 x_{ij} + (u_{0j} + u_{1j} x_{1ij} + e_{0ij} + e_{1ij} x_{1ij}), \tag{3.6}$$

$$\text{var}(e_{0ij}) = \sigma_{e0}^2, \quad \text{var}(e_{1ij}) = 0, \quad \text{cov}(e_{0ij} e_{1ij}) = \sigma_{e01}$$

the level 1 variance is now the linear function $\sigma_{e0}^2 + 2\sigma_{e01} x_{1ij}$.

This device of constraining a variance parameter to be zero in the presence of a non-zero covariance is used to obtain the required variance structure. Thus it is only the specified function of the random parameters in expressions such as (3.6) which have an interpretation in terms of the level 1 variances of the responses y_{ij}. Furthermore, we can allow a parameter such as σ_{e1}^2 to be negative, so long as the total level 1 variance remains positive within the range of the data. For convenience, however, we will continue to use the notation $\text{var}(e_{1ij}) = \sigma_{e1}^2$ etc. As we have already seen, model A shows a clear quadratic relationship which is highly statistically significant (chi-squared with 2 degrees of freedom = 123). Furthermore, the level 2 correlation between the intercept and slope is now reduced to -0.91 from the value of -1.03 from Table 2.5, and with little change among the fixed part coefficients.

This procedure can be used generally at any level to define complex variation where the random coefficients vary at the same level at which the explanatory variables are defined. Thus for example, in the analyses of the JSP data in Chapter 2, we could model the average school 8-year score, which is a level 2 variable, as having a random coefficient at level 2. If the resulting variance and covariance are non-zero, the interpretation will be that the between-school variance is a quadratic function of the 8-year score, namely $\sigma_{u0}^2 + 2\sigma_{u01}x_{1j} + \sigma_{u1}^2 x_{1j}^2$ where x_{1j} is the average 8-year score.

The model (3.6) does not constrain the overall level 1 contribution to the variance in any way. In fact, if we try to fit this model to the JSP data we run into convergence problems since the linear function estimated at each iteration sometimes becomes negative within the range of the data. We shall see in Chapter 8 how the alternative nonlinear modelling of the variance overcomes this problem. If we use MCMC estimation we can overcome this problem by constraining the level 1 variance to be positive at each iteration, as follows.

Consider (3.5) where we have

$$\text{var}(e_{ij}) = \text{var}(e_{0ij} + e_{1ij}x_{1ij}) = \sigma_{e0}^2 + 2\sigma_{e01}x_{1ij} + \sigma_{e1}^2 x_{1ij}^2 \qquad (3.7)$$

At a given iteration t, suppose that we wish to update the parameter σ_{e01} given current values. Since the expression (3.7) is to remain positive for all observed values of x_{1ij} we require

$$\sigma_{e01} + \frac{\sigma_{e0}^2 + \sigma_{e1}^2 x_{1ij}^2}{2x_{1ij}} > 0$$

We can exclude the case $x_{1ij} = 0$, since then the variance is simply σ_{e0}^2. For the negative values of x_{1ij} we require

$$\sigma_{e01} < \min\left(\frac{\sigma_{e0}^2 + \sigma_{e1}^2 x_{1ij}^2}{-2x_{1ij}}\right)$$

and for positive values we require

$$\sigma_{e01} > \max\left(\frac{\sigma_{e0}^2 + \sigma_{e1}^2 x_{1ij}^2}{-2x_{1ij}}\right)$$

These constraints define a range of admissible values. Thus a standard MH proposal distribution can be used with these constraints defining truncation points for the distribution at each iteration. Each of the level 1 variance parameters is updated similarly, although for these cases only one truncation point is needed. The other parameters can be updated using Gibbs steps. Browne (2002) describes the details for a Normal proposal distribution.

One of the reasons for the high negative correlation between the intercept and slope at the school level may be associated with the fact that the 11-year score has a 'ceiling' with a third of the students having scores of 35 or more out of 40. A standard procedure for dealing with such skewed distributions is to transform the

Table 3.1   JSP mathematics data with a level 1 variance a quadratic function of 8-year score measured about the sample mean. Model A with original scale; models B and C with Normal score transform of 11-year score and original 8-year score measured about its mean

Parameter	Estimate (s.e.)		
	A	B	C
Fixed			
Intercept (x_0)	31.2	0.13	0.14
8-year score (x_1)	0.99 (0.29)	0.097 (0.004)	0.096 (0.004)
Gender (boys–girls)	−0.35 (0.26)	−0.04 (0.05)	−0.03 (0.05)
Social class (non-manual–manual)	0.74 (0.29)	0.16 (0.06)	0.16 (0.06)
School mean 8-year score	0.02 (0.11)	−0.008 (0.02)	
8-year score × school mean 8-year score	−0.02 (0.01)	0.0006 (0.02)	
Random			
Level 2			
σ_{u0}^2	2.84 (0.88)	0.084 (0.024)	0.086 (0.024)
σ_{u01}	−0.17 (0.07)	−0.0024 (0.0015)	−0.0030 (0.0015)
σ_{u1}^2	0.012 (0.007)	0.00018 (0.00016)	0.00021 (0.00016)
Level 1			
σ_{e0}^2	16.5 (1.02)	0.413 (0.029)	0.412 (0.022)
σ_{e01}	−0.90 (0.02)	−0.0032 (0.0017)	
σ_{e1}^2	0.06 (0.02)	0.0000093 (0.00041)	

data, for example to Normality, and this is most conveniently done by computing Normal scores; that is by assigning Normal order statistics to the ranked scores as described in 2.9, and we shall adopt this procedure as standard in future examples. The results from this analysis are given under model B in Table 3.1. Note that the scale has changed since the response is now a standard normal variable with zero mean and unit standard deviation. We now find that there is no longer any appreciable complex variation at level 1; the chi-squared test yields a value of 3.4 on 2 degrees of freedom. Nor is there any effect of the compositional variable of mean school 8-year score; the chi-squared test for the two fixed coefficients associated with this give a value of 0.2 on 2 degrees of freedom. The reduced model is fitted as C. The parameters associated with the random slope at level 2 remain significant ($\chi_2^2 = 7.7$, $P = 0.02$) and the level 2 correlation is further reduced to −0.71. Figure 3.3 shows the level 1 standardized residuals plotted against the predicted values from which it is clear that now the variance is much more nearly constant. This example demonstrates that interpretations may be sensitive to the scale on which variables are measured. It is typical of many measurements in the social sciences that their scales are arbitrary and we can justify nonlinear, but monotone, order preserving, transformations if they help to simplify the statistical model and the interpretation. In section 2.10 we also showed how models which used different sets of *explanatory* variables, not necessarily nested ones, could be compared using the AIC criterion. Together with studying the effects of transformations of the response this can be used to arrive at a final model that provides an optimum fit to the data.

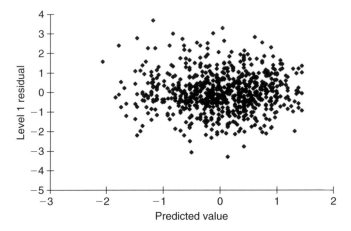

Figure 3.3   Level 1 standardized residuals by predicted values for analysis C in Table 3.1.

We are not limited to making the variance a function of explanatory variables that appear in the fixed part of the model. A traditional, single-level, example is 'regression through the origin' in which the fixed intercept term is zero while a level 1 variance associated with the intercept is fitted.

We can consider any particular function of explanatory variables as the basis for modelling the variance. One possibility is to take the fixed part predicted value $\hat{y}_{ij}$ and define the level 1 random term as $e_{1ij}\sqrt{\hat{y}_{ij}}$, assuming the predicted value is positive, so that the level 1 variance becomes $\sigma_{e1}^2\hat{y}_{ij}$, that is proportional to the predicted value; often known as a 'constant coefficient of variation' model that we will discuss later.

3.1.1   Partitioning the variance and intra-unit correlation

In Chapter 2 we introduced the variance partition coefficient (VPC) for a 2-level variance component model defined as

$$\tau = \frac{\sigma_u^2}{\sigma_u^2 + \sigma_e^2}$$

which is the proportion of the total variance occurring at level 2. For this model this is also the correlation between two level 1 units in the same level 2 unit (the intra-unit correlation ρ). In more complex models such as (3.5), however, the intra-unit correlation is not equal to the VPC. Thus for (3.5) the correlation between two students (i_1, i_2) in the same school with scores x_{1i_1j}, x_{1i_2j} is

$$\frac{(\sigma_{u0}^2 + \sigma_{u01}(x_{1i_1j} + x_{1i_2j}) + \sigma_{u1}^2 x_{1i_1j} x_{1i_2j})}{(\sigma_{u0}^2 + 2\sigma_{u01}x_{1i_1j} + \sigma_{u1}^2 x_{1i_1j}^2 + \sigma_{e0}^2 + 2\sigma_{e01}x_{1i_1j} + \sigma_{e1}^2 x_{1i_1j}^2)^{1/2}}$$

$$\times \frac{1}{(\sigma_{u0}^2 + 2\sigma_{u01}x_{1i_2j} + \sigma_{u1}^2 x_{1i_2j}^2 + \sigma_{e0}^2 + 2\sigma_{e01}x_{1i_1j} + \sigma_{e1}^2 x_{1i_1j}^2)^{1/2}}$$

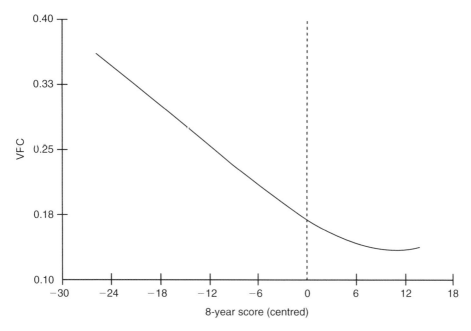

Figure 3.4 Variance partition coefficient (VPC) for analysis B in Table 3.1.

whereas the VPC values are

$$\frac{(\sigma_{u0}^2 + 2\sigma_{u01}x_{1i_1j} + \sigma_{u1}^2 x_{1i_1j}^2)}{(\sigma_{u0}^2 + 2\sigma_{u01}x_{1i_1j} + \sigma_{u1}^2 x_{1i_1j}^2 + \sigma_{e0}^2 + 2\sigma_{e01}x_{1i_1j} + \sigma_{e1}^2 x_{1i_1j}^2)} \quad \text{and}$$

$$\frac{(\sigma_{u0}^2 + 2\sigma_{u01}x_{1i_2j} + \sigma_{u1}^2 x_{1i_2j}^2)}{(\sigma_{u0}^2 + 2\sigma_{u01}x_{1i_2j} + \sigma_{u1}^2 x_{1i_2j}^2 + \sigma_{e0}^2 + 2\sigma_{e01}x_{1i_2j} + \sigma_{e1}^2 x_{1i_2j}^2)}$$

Figure 3.4 shows the VPC as a function of the 8-year score for analysis B in Table 3.1. The relative variation at the school level decreases steadily with increasing 8-year score. A tentative interpretation is that the school attended is of more importance for those with below average prior achievement.

In Chapter 4 we shall consider the calculation of the VPC for models with discrete responses.

3.1.2 Variances for subgroups defined at level 1

A common example of complex variation at level 1 is where variances are specific to subgroups. For example, for many measurements there are gender or social class differences in the level 1 variation. A straightforward way to model this situation in the case of a single such grouping is by defining the following version of (3.5) for a

Table 3.2   JSP data with normal score of 11-year maths as response. Subscript 1 refers to 8-year maths score, 2 to manual group, 3 to non-manual group, 4 to boys

Parameter	Estimate (s.e.)		
	A	B	C
Fixed			
Constant	0.13	0.13	0.13
8-year score	0.096 (0.004)	0.096 (0.004)	0.096 (0.004)
Gender (boys–girls)	−0.03 (0.05)	−0.03 (0.05)	−0.03 (0.05)
Social class	0.16 (0.05)	0.16 (0.05)	0.16 (0.05)
(non-manual–manual)			
Random			
Level 2			
σ^2_{u0}	0.086 (0.025)	0.086 (0.025)	0.086 (0.024)
σ_{u01}	−0.0029 (0.0015)	−0.0029 (0.0015)	−0.0028 (0.0015)
σ^2_{u1}	0.00018 (0.00015)	0.00018 (0.00015)	0.00018 (0.00015)
Level 1			
σ^2_{e0}		0.37 (0.04)	0.36 (0.04)
σ_{e02}		0.03 (0.02)	0.03 (0.02)
σ^2_{e2}	0.43 (0.03)		
σ^2_{e3}	0.37 (0.04)		
σ_{e04}			0.004 (0.02)
−2 (log-likelihood)	1491.8	1491.8	1491.7

model with different variances for children with manual and with non-manual social class backgrounds.

$$y_{ij} = \beta_0 + \beta_1 x_{ij} + (u_{0j} + e_{2ij}z_{2ij} + e_{3ij}z_{3ij})$$

$$z_{2ij} = 1 \text{ for manual,} \quad 0 \text{ for non-manual}$$

$$z_{3ij} = 0 \text{ for manual,} \quad 1 \text{ for non-manual}$$

$$\text{var}(e_{2ij}) = \sigma^2_{e2} \quad \text{var}(e_{3ij}) = \sigma^2_{e3} \quad \text{cov}(e_{2ij}, e_{3ij}) = 0$$

If we do this for model C in Table 3.1 then we obtain the estimates in column A of Table 3.2. The estimates of the fixed parameters have changed little and the level 2 parameters are also similar. At level 1 the variance for the manual students is higher than that for the non-manual students, but not significantly so since the likelihood ratio test statistic, formed by differencing the values of (−2 log-likelihood) for the model with a single-level 1 variance (1493.7) and that given in analysis A of Table 3.2, gives a chi-squared test statistic of 1.9 on 1 degree of freedom.

We now look at an alternative method for specifying this type of complex variation at level 1 which has certain advantages. We write

$$y_{ij} = \beta_0 + \beta_1 x_{ij} + (u_{0j} + e_{2ij}z_{2ij})$$

$$z_{2ij} = 1 \text{ for manual,} \quad 0 \text{ for non-manual}$$

$$\text{var}(e_{0ij}) = \sigma^2_{e0}, \quad \text{var}(e_{2ij}) = 0, \quad \text{cov}(e_{0ij}, e_{2ij}) = \sigma_{e02}$$

and the level 1 variance is given by $\sigma^2_{e0} + 2\sigma_{e02}z_{2ij}$ because we have constrained the variance of the manual coefficient to be zero. Thus, for manual children ($z_{2ij} = 1$) the level 1 variance is $\sigma_{e0}^2 + 2\sigma_{e02}$ and for non-manual children the level 1 variance is σ_{e0}^2. The second column in Table 3.2 gives the results from this formulation and we see that, as expected, the covariance estimate is equal to half the difference between the separate variance estimates in the first column.

Suppose now that we wish to model the level 1 variance as a function both of social class group and gender. One possibility is to fit a separate variance for each of the four possible resulting groups, using either of the above procedures. Another possibility is to consider a more parsimonious 'additive' model for the variances as follows

$$e_{ij} = e_{0ij} + e_{2ij}z_{2ij} + e_{4ij}z_{4ij}$$

$$z_{4ij} = 1 \text{ if a boy, } 0 \text{ if a girl}$$

$$\text{var}(e_{0ij}) = \sigma_{e0}^2, \quad \text{cov}(e_{0ij}e_{2ij}) = \sigma_{e02}, \quad \text{cov}(e_{0ij}e_{4ij}) = \sigma_{e04} \qquad (3.8)$$

with the remaining two variances and covariance equal to zero. Thus (3.8) implies that the level 1 variance for a manual boy is $\sigma_{e0}^2 + 2\sigma_{e02} + 2\sigma_{e04}$ etc. The third column of Table 3.2 gives the estimates for this model and we see that there is a negligible difference in the level 1 variance for boys and girls.

We can extend such structuring to the case of multicategory variables and we can also include continuous variables as in Table 3.1. Suppose we had a three-category variable: we define two dummy variables, say z_{5ij}, z_{6ij} corresponding to the second and third categories, just as if we were fitting the factor in the fixed part of the model. With z_{1ij} representing the continuous variable an additive model for the level 1 random variation can be written as

$$e_{ij} = e_{0ij} + e_{1ij}z_{1ij} + e_{5ij}z_{5ij} + e_{6ij}z_{6ij}$$

$$\text{var}(e_{0ij}) = \sigma_{e0}^2, \quad \text{var}(e_{1ij}) = \sigma_{e1}^2, \quad \text{cov}(e_{0ij}e_{1ij}) = \sigma_{e01}$$

$$\text{cov}(e_{0ij}e_{5ij}) = \sigma_{e05}, \quad \text{cov}(e_{0ij}e_{6ij}) = \sigma_{e06}$$

with the remaining terms equal to zero.

This model can be elaborated by including one or both the covariances between the dummy variable coefficients and the continuous variable coefficient, namely σ_{e15}, σ_{e16}. These covariances are analogous to interaction terms in the fixed part of the model and we see that, starting with an additive model, we can build up models of increasing complexity. The only restriction is that we cannot fit covariances between the dummy variable categories for a single explanatory variable. Thus if social class had three categories, we could fit two covariances corresponding to, say, categories 2 and 3 but not a covariance *between* these categories.

Residuals can be estimated in a straightforward manner for these complex variation models. For example, from (3.8) the estimated residual for a manual boy is $\hat{e}_{0ij} + \hat{e}_{2ij} + \hat{e}_{4ij}$ where the estimates of the individual residuals are computed using the formulae in Appendix 2.2 with the appropriate zero variances.

3.1.3   Variance as a function of predicted value

The level 1 variance can be modelled as a function of any combination of explanatory variables and in particular we can incorporate the estimated coefficients themselves in

Table 3.3   GCSE scores related to secondary school intake achievement

Parameter	Estimate (s.e.)	
	A	B
Fixed		
Constant	0.13	0.14
Reading score	0.50 (0.03)	0.49 (0.03)
Gender (boys–girls)	−0.19 (0.06)	−0.22 (0.06)
Social class (non-manual–manual)	−0.07 (0.06)	−0.06 (0.06)
Random		
Level 2		
σ^2_{u0}	0.03 (0.02)	0.02 (0.01)
Level 1		
σ^2_{e0}	0.66 (0.04)	0.63 (0.04)
σ_{e01}		0.16 (0.04)
σ^2_{e1}		0.11 (0.09)
−2 (log-likelihood)	1929.5	1905.0

such functions. A useful special case is where the function is the fixed part predicted value $\hat{y}_{ij}$. Thus (3.7) becomes

$$y_{ij} = \beta_0 + \beta_1 x_{ij} + (u_{0j} + e_{0ij} + e_{1ij}\hat{y}_{ij})$$

with level 1 variance given by $\sigma^2_{e0} + 2\sigma_{e01}\hat{y}_{ij} + \sigma^2_{e1}\hat{y}^2_{ij}$. A special case of this model is the so-called 'constant coefficient of variation model' where the two variance terms are constrained to zero, and $\hat{y}_{ij}$ remains positive. The estimation of the random parameters is straightforward: at each iteration of the algorithm a new set of predicted values are calculated and used as the level 1 explanatory variable.

Table 3.3 illustrates the use of this model where the level 1 variance shows a strong dependence on the predicted value. The data are the General Certificate of Secondary Examination (GCSE) scores at the age of 16 years of the Junior School Project students. This score is derived by assigning values to the grades achieved in each subject examination and summing these to produce a total score (see Nuttall et al. (1989) for a detailed description). There are 785 students in this analysis in 116 secondary schools to which they transferred at the age of 11 years. The students have a measure of reading achievement, the London Reading Test (LRT) taken at the end of their junior school, and this is used as a pretest baseline measure against which relative progress is judged. Both the reading test score and the examination score have been transformed to Normal equivalent deviates.

Analysis A is a variance components analysis and Figure 3.5 shows a plot of the standardized level 1 residuals against the predicted values. It is clear that the variation is much smaller for low predicted values.

One possible extension of the model to deal with this is to use the LRT score as an explanatory variable at level 1, so that the level 1 variance becomes a quadratic function of LRT score. This does not, however, entirely eliminate the relationship and instead we model the predicted value as a level 1 explanatory variable, and the results are presented as analysis B of Table 3.3. If we now plot the standardized residuals associated with the intercept against the predicted values we obtain the pattern in Figure 3.6, from which

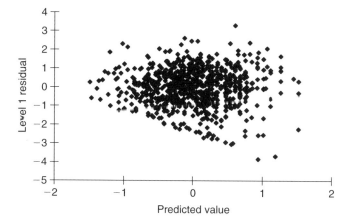

Figure 3.5 Standardized residuals for variance components analysis in Table 3.3.

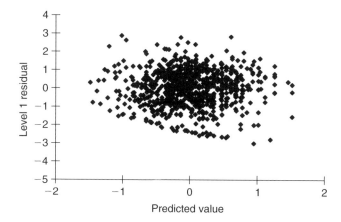

Figure 3.6 Standardized residuals with level 1 variance as a function of predicted value in Table 3.3. Star symbol indicates school average.

it is clear that much of the relationship between the variance and the predicted value has been accounted for. We could go on to fit more complex functions of the predicted value, for example involving nonlinear or higher order polynomial terms.

3.1.4 Variances for subgroups defined at higher levels

The random slopes model in Table 3.1 has already introduced complex variation at level 2 when the coefficient of a level 1 explanatory variable is allowed to vary across level 2 units. Just as with level 1 complex variation, we can also allow coefficients of variables defined at level 2 to vary at level 2. Exactly the same considerations apply for categorical level 2 variables as we had for such variables at level 1 and complex additive or interactive structures can be defined.

The coefficient of a level 2 variable can vary randomly at either level 1 or level 2 or both. For example, suppose we have three types of school: all boys schools, all girls schools and mixed schools. We can allow different variances, at level 2, between boys' schools, between girls' schools and between mixed schools. We can also allow different between-student variances for each type of school.

To further illustrate complex level 1 variation and also to introduce a 3-level model we look at another dataset, this time from a survey of social attitudes.

3.2   A 3-level complex variation model

The longitudinal or panel data come from the British Social Attitudes Survey and cover the years 1983 to 1986 with a random sample of 264 adults measured a year apart on four occasions and living at the same address. This panel was a subsample of a larger series of cross-sectional surveys. The final sample was intended to be self weighting with each household as represented by a single person having the same inclusion probability. A full technical account of the sampling procedures is given by McGrath and Waterton (1986). The sampling procedure was at the first stage to sample parliamentary constituencies (the primary sampling units – PSUs) with probability proportional to size of electorate, then to sample a single 'polling district' within each constituency in a similar way and finally to sample an equal number of addresses within each polling district.

Because only one polling district was sampled from each constituency, we cannot separate the between-district from the between-constituency variation; the two are 'confounded'. Likewise we cannot separate the between-individual from the between-household variation. The basic variation is therefore at two levels, between districts (constituencies) and between individuals (households). The longitudinal structure of the data, with four occasions, introduces a further level below these two, namely a between-occasion-within-individual level, so that occasion is level 1, individual is level 2 and district is level 3. In Chapter 5 we shall study longitudinal data structures in more depth, both at level 1 and higher levels.

The response variable we shall use is a scale, in the range 0–7, concerned with attitudes to abortion. It is derived by summing the (0, 1) responses to seven questions and can be interpreted as indicating whether the respondent supported or opposed a woman's right to abortion with high scores indicating strong support. Explanatory variables are political party allegiance (four categories), self-assessed social class (three categories), gender, age (continuous), and religion (four categories) and year (four categories). A number of preliminary analyses have been carried out and the effects of party allegiance, social class, gender, and age, were found to be small and not statistically significant. We therefore examine the basic 3-level model which can be written as follows:

$$y_{ijk} = \beta_0 + (\beta_1 x_{1ijk} + \beta_2 x_{2ijk} + \beta_3 x_{3ijk})$$
$$+ (\beta_4 x_{4ijk} + \beta_5 x_{5ijk} + \beta_6 x_{6ijk}) + (v_k + u_{jk} + e_{ijk}) \qquad (3.9)$$

with the explanatory variables with subscripts 1–3 being dummy variables for religious categories 2–4 and those with subscripts 4–6 being dummy variables for years 1984 to 1986. We have three variances, one at each level in the random part of the model. The response variable in the following analyses has only eight categories, with 32% of the sample having the highest value of 7. The response has been transformed by assigning

Table 3.4 Repeated measurements of attitudes to abortion. Response is Normal score transformation. Religion estimates are contrasted with none; age is measured about the mean of 37 years

Parameter	Estimate (s.e.)		
	A	B	C
Fixed			
Constant	0.32	0.33	0.33
Religion			
Roman Catholic	−0.80 (0.18)	−0.80 (0.18)	−0.69 (0.18)
Protestant	−0.27 (0.10)	−0.26 (0.10)	−0.25 (0.10)
Other	−0.63 (0.13)	−0.63 (0.13)	−0.54 (0.14)
Year			
1984	−0.29 (0.05)	−0.29 (0.05)	−0.29 (0.05)
1985	−0.06 (0.05)	−0.07 (0.05)	−0.07 (0.05)
1986	0.06 (0.05)	0.05 (0.04)	0.05 (0.04)
Age			0.013 (0.005)
Age × Roman Catholic			−0.036 (0.010)
Age × Protestant			−0.014 (0.007)
Age × Other			−0.023 (0.008)
Random			
Level 3			
σ_v^2	0.03 (0.03)	0.03 (0.02)	0.03 (0.02)
Level 2			
σ_u^2	0.37 (0.04)		0.34 (0.04)
Level 1			
σ_{e0}^2	0.31 (0.02)	0.21 (0.08)	0.21 (0.03)
σ_{e01}		0.11 (0.05)	0.10 (0.04)
σ_{e02}		0.03 (0.16)	0.03 (0.02)
σ_{e03}		0.04 (0.02)	0.04 (0.02)
σ_{e04}		0.05 (0.02)	0.05 (0.02)
σ_{e05}		0.05 (0.02)	0.05 (0.02)
σ_{e06}		0.00 (0.02)	0.00 (0.02)
−2 (log-likelihood)	2233.5	2214.2	2198.7

Normal scores to the overall distribution and we shall treat the response as if it was continuously distributed. In Chapter 4 we shall look at other models which retain the categorization of the response variable.

Table 3.4 gives the results of fitting (3.9). The between-occasion and between-individual variances are similar. The level 3 variance is small, and the likelihood ratio chi-squared is 2.05 (compared with a value of 1.64 obtained from comparing the estimate with its standard error), which is not significant at the 10% level.

For the religious differences we have $\chi_3^2 = 33.7$ for the overall test with all those having religious beliefs being less inclined to support abortion, the Roman Catholic and Other religions being least likely of all. The Roman Catholic and Other religions are significantly less likely than the Protestants to support abortion. The simultaneous test (3 d.f.) chi-squared statistics respectively are 9.7 and 9.0 ($P = 0.03$).

For the year differences we have $\chi_3^2 = 59.7$ and simultaneous comparisons show that in 1984 there was a substantially less approving attitude towards abortion. It is likely that this is an artefact of the way questions were put to respondents.[1] No significant interaction exists between religion and year.

We now look at elaborating the random structure of the model. At level 1 we fit an additive model as in section 3.1.1 for the categories of religion and for year. Year is the variable defining level 1, but religion is defined at level 2 and is an example of a higher level variable used to define complex variation at a lower level.

The results are given as analysis B in Table 3.4. For year we obtain $\chi_3^2 = 8.3$ ($P = 0.04$) and for religion $\chi_3^2 = 11.0$ ($P = 0.01$). There is a greater heterogeneity within the Roman Catholics, from year to year, and within the Other religions than within Protestants and those with no religion. The addition of these variances to the model does not change substantially the values for the other parameters.

Fitting complex variation at level 2 (between individuals) and level 3 (between districts) does not yield statistically significant effects, although there is some suggestion that there may be more variation among Roman Catholics.

For the final analysis we look again at the fixed part and explore interactions. None of the interactions have important effects except for that of age with religion, although age on its own had a negligible effect. We see from analysis C that those with no religion show an increasing approval of abortion with age, whereas the Roman Catholics and to a smaller extent Other religions show a decreasing approval with age. The overall chi-squared for testing the interactions is 16.1 with 3 degrees of freedom ($P = 0.001$).

3.3   Parameter constraints

In the example of the previous section some of the fixed and random parameters for year and religious groups were similar. This suggests that we could fit a simpler model by forcing or 'constraining' such parameters to take the same values and also thus decreasing the standard errors in the model. We illustrate the procedure using the fixed part estimates for the abortion attitudes data.

We consider the general constraint for the fixed parameters in the form of n linear constraint functions defined by $C^T\beta = k$, where C is a $(p \times n)$ constraint matrix and k is a vector, which can have quite general values for their elements.

Suppose that, in analysis C of Table 3.4, we wished to constrain the main effects and interaction terms of the Roman Catholic and Other religions to be equal. This implies $n = 2$ constraint functions, and we have

$$C^T = \begin{pmatrix} 0 & 1 & 0 & -1 & 0 & 0 & 0 & 0 & 0 & 0 & 0 \\ 0 & 0 & 0 & 0 & 0 & 0 & 0 & 0 & 1 & 0 & -1 \end{pmatrix}$$

$$k = \begin{pmatrix} 0 \\ 0 \end{pmatrix}$$

which implies $\hat{\beta}_1 = \hat{\beta}_3, \hat{\beta}_8 = \hat{\beta}_{10}$.

---

[1] In 1984 seven questions making up the attitude scale were put to respondents in the reverse order, that is with the most 'acceptable' reasons for having an abortion (e.g. as a result of rape) coming first. This illustrates an important issue in surveys of all kinds which collect data for comparisons over time, namely to maintain the same questioning procedure.

The constrained estimator of β is

$$\hat{\beta}^c = \hat{\beta} - LC(C^T LC)^{-1}(C^T \hat{\beta} - k)$$
$$L = (X^T \hat{V}^{-1} X)^{-1}$$

(3.10)

where $\hat{\beta}$ is the unconstrained estimator. The covariance matrix of the constrained estimator is MLM^T where

$$M = I - LC(C^T LC)^{-1} C^T$$

There is an analogous formula for constrained random parameters. Using the above constraints for analysis C in Table 3.4, the random parameters are little changed, the main effects for Roman Catholic and Other religions become -0.57 and the interaction terms become -0.026 and the remaining main effects are virtually unaltered. The standard errors, as expected, are smaller, being 0.121 for the main effect estimate and 0.007 for the interaction.

In addition to linear constraints we can also apply nonlinear constraints. To illustrate the procedure we consider the analysis in Table 2.5, where the estimated correlation between the slope and intercept was -1.03. To constrain this to be exactly -1.0, after each iteration of the algorithm we compute the covariance as a function of the variances to give this correlation. Thus, after iteration t we compute $\sigma_{u01}^{t+1} = \hat{\sigma}_{u0}^t \hat{\sigma}_{u1}^t$ and then constrain the covariance to be equal to this value, a linear constraint, for iteration $t + 1$. This procedure is repeated until convergence is obtained for the unconstrained values. For more general nonlinear constraints we may require several such constraints to apply simultaneously.

If we constrain the model of Table 2.5 to give a correlation of -1.0 we find that the fixed effects and the level 1 variance are altered only slightly, with a small reduction in standard errors. The level 2 parameters, however, are reduced by about 50% and are closer to those in analysis A of Table 3.1 where the estimated correlation is -0.91.

We can also temporarily constrain values during the iterative estimation procedure if convergence is difficult or slow. Some parameters, or functions of them, can be held at current values, other parameter values allowed to converge and the constrained parameters subsequently unconstrained.

3.4 Weighting units

It is common in sample surveys to select level 1 units, for example household members, so that each unit in the population has the same probability of selection. Such self-weighting samples can then be modelled using any of the multilevel models of this book. Likewise, if the model correctly specifies the population structure, non-self-weighting samples can be modelled similarly: the differential selection probabilities contain no extra information for the model parameters. If we wished to form predictions for the whole population, or well-defined sub-population, on the basis of the model estimates, we could apply the weight (typically the inverse of the selection probability) to the predicted value for each level 1 unit and then form a weighted sum over these units. Chapter 9 further discusses estimation from sample surveys.

In some cases, however, the sample inclusion probabilities are designed to be equal but there is differential non-response. It is common in such cases to assign weights to the level 1 units to compensate for the non-response. Another situation where weighting should be used is when the modelling of the population structure is inadequate, that is the model is misspecified. If we then wish to make inferences about population values which are robust against poor model specification we will need to weight the units. In some data analyses, we may come across values which are possible errors. Rather than excluding the units containing these, we may wish to keep them but assign them a lower weighting in the analysis in line with the extent to which we believe they are in error.

In Chapter 9 we shall investigate in detail the application of weighting for sample surveys. In particular we shall look at the case where weighting is used to compensate for non-response and the probability of responding is 'informative', that is associated with the response variable of interest. In the remainder of this section we will describe the application of weights under the assumption that the weights themselves are unrelated to the random effects. In fact, Pfefferman et al. (1997) conclude that, even where weights are informative, the following procedure produces acceptable results in many cases but can give biased results in some circumstances and should be used with caution.

Consider the case of a 2-level model. Denote by w_j the weight attached to the j-th level 2 unit and by $w_{i|j}$ the weight attached to the i-th level 1 unit within the j-th level 2 unit. This allows for the possibility of having differential weights at both levels. We scale the weights so that

$$\sum_i w_{i|j} = n_j, \quad \sum_j w_j = m \tag{3.11}$$

where m is the total number of level 2 units and $N = \sum_j n_j$ the total number of level 1 units. That is, the lower level weights within each immediate higher level unit are scaled to have a mean of unity, and likewise for higher levels. For each level 1 unit we now form the final, or composite, weight

$$w_{ij} = N w_{i|j} w_j \Big/ \sum_{i,j} w_{i|j} w_j = N w_{i|j} w_j \Big/ \sum_j n_j w_j \tag{3.12}$$

Denote by Z_u, Z_e respectively the sets of explanatory variables defining the level 2 and level 1 random coefficients and form

$$Z_u^* = W_j Z_u, \quad W_j = diag\{w_j^{-0.5}\}$$
$$Z_e^* = W_{ij} Z_e, \quad W_{ij} = diag\{w_{ij}^{-0.5}\} \tag{3.13}$$

We now carry out a standard estimation but using Z_u^*, Z_e^* as the random coefficient explanatory variables. For a 3-level model, with an obvious extension to notation, we have the following:

$$\sum_i w_{i|jk} = n_{jk}, \quad \sum_j w_{j|k} = m_k, \quad \sum_k w_k = K, \quad N = \sum_{jk} n_{jk}, \quad m = \sum_k m_k$$

$$w_{ijk} = N w_{i|jk} w_{j|k} w_k \Big/ \sum_{ijk} w_{i|jk} w_{j|k} w_k, \quad w_{jk} = m w_{j|k} w_k \Big/ \sum_{jk} w_{j|k} w_k$$

where K is the number of level 3 units.

Denote by V^* the weighting matrix in this analysis. The fixed part coefficient estimates and their covariance matrix are given by

$$\hat{\beta} = (X^T V^{*-1} X)^{-1} X^T V^{*-1} Y,$$
$$\text{cov}(\hat{\beta}) = (X^T V^{*-1} X)^{-1} X^T V^{*-1} V V^{*-1} X (X^T V^{*-1} X)^{-1} \quad (3.14)$$

with an analogous result for the random parameter estimates. The covariance matrix for the random parameters is complex, but a sandwich estimator can be used (see section 3.5).

In survey work analysts often have access only to the final level 1 weights w_{ij}. In this case, say for a 2-level model, we can obtain the w_j by computing $w'_j = m W_j / \Sigma_j W_j$, $W_j = (\Sigma_i w_{ij})/n_j$. For a 3-level model the procedure is carried out for each level 3 unit and the resulting w'_{jk} are transformed analogously.

A number of features are worth noting. Firstly, for a single-level model this procedure gives the usual weighted regression estimator. Secondly, suppose we set a particular level 1 weight to zero. This is not equivalent to removing that unit from the analysis in a 2-level model since the level 2 (weighted) contribution remains. Nevertheless, this weighting may be appropriate if we wish to remove the effect of the unit only at level 1, say if it were an extreme level 1 outlier. If, however, we set a level 2 weight to zero then this is equivalent to removing the complete level 2 unit. If we wished to obtain estimates equivalent to removing the level 1 unit we would need to set all the level 2 (random coefficient) explanatory variables for that level 1 unit to zero also. This is easily done by defining an indicator variable for the unit (or units) with a zero corresponding to the unit in question and multiplying all the random explanatory variables by it.

3.4.1 Weighted residuals

For the level 2 residuals with weights attached to the responses the residual estimates (see Appendix 2.2) are

$$\hat{u}^{(2)} = \Omega_2 Z_u^T V^{-1} W_{ij}^{-1} \tilde{Y},$$
$$\text{cov}(\hat{u}^{(2)}) = \Omega_2 Z_u^T V^{-1} W_{ij}^{-1} V W_{ij}^{-1} V^{-1} Z_u \Omega_2 \quad (3.15)$$
$$V = E(\tilde{Y}\tilde{Y}^T)$$

where $W_{ij}^{-1} = diag(w_{ij})$, the final level 1 weights.

This provides a consistent estimator of the diagnostic covariance matrix. The comparative covariance matrix is given by

$$E[(\hat{u}^{(2)} - u^{(2)})(\hat{u}^{(2)} - u^{(2)})^T] = \Omega_2 Z_u^T V^{-1} W_{ij}^{-1} V W_{ij}^{-1} V^{-1} Z_u^T \Omega_2$$
$$- 2\Omega_2 Z_u^T V^{-1} W_{ij}^{-1} Z_u \Omega_2 + \Omega_2$$

For many purposes an equally weighted estimator for the residuals is adequate, in which case the usual formulae apply.

Note that in principle we could choose any weighting matrix for the responses in (3.15). For example, in a 2-level model weights may be applied in the estimation to reflect an overall sampling procedure, but we may wish to apply different weights to reflect the composition of each level 2 unit separately. For example, suppose we have schools with an equally weighted sample where every level 1 unit has the same probability of selection. The actual sample proportions of, say, male and female students may be very different from that in the school as a whole. If there are important gender differences, then in such a case we may wish to apply weights which reflect the true rather than the sample proportions.

A similar procedure applies for multilevel generalized linear models which are discussed in Chapter 4. Here the weighted explanatory variables at levels 2 and higher are as above. For the quasilikelihood estimators (PQL and MQL) at level 1 the vector Z_e is that which defines the binomial variation. Thus, for binomial data, at level 1 a method of incorporating the weight vector is to use Z_e but to work with $w_{ij}n_{ij}$ instead of n_{ij} as the denominator.

3.5   Robust (sandwich) estimators and jackknifing

Until now we have assumed that the response variable has a Normal distribution, and where the departure from Normality is substantial we have considered a transformation, using Normal scores. As we saw in the abortion dataset, however, such transformations may be only approximate where the original score distribution is highly discrete or very skew. The estimates of the fixed and random parameter estimates will still be consistent when the Normality assumption is untrue, but the standard error estimates cannot be used to obtain confidence intervals or to test significance except in large samples. In other cases we may have misspecified the random part of the model.

One way of attempting to deal with this problem is to develop estimators which are based upon alternative distributional assumptions, and in later chapters we shall adopt this approach when dealing with discrete and ordered response data. We have also seen in Chapter 2 the use of a t-distribution as an alternative to the Normal for the level 1 residuals. Seltzer (1993) gives another example, using Gibbs sampling, with a t-distribution.

An alternative procedure is to modify the standard error and confidence interval estimates so that they are less dependent on distributional assumptions, of whatever kind. One of the penalties of this is that the resulting significance tests and confidence intervals will tend to be wider, or more 'conservative', than those derived under a particular distributional assumption.

Consider first the fixed part of the model and the usual IGLS estimate of the fixed parameters based upon the random parameter estimates

$$\hat{\beta} = (X^T \hat{V}^{-1} X)^{-1} X^T \hat{V}^{-1} Y$$

The covariance matrix of these estimates is

$$\text{cov}(\hat{\beta}) = (X^T \hat{V}^{-1} X)^{-1} X^T \hat{V}^{-1} \{\text{cov}(Y)\} \hat{V}^{-1} X (X^T \hat{V}^{-1} X)^{-1} \qquad (3.16)$$

where $\text{cov}(Y) = V$ and is unknown because we cannot rely upon the usual Normality assumption when estimating V (see Appendix 2.1) or because the random part of the model is misspecified. The usual procedure would be to substitute the estimated $\hat{V}$, but this will generally lead to standard errors which are too small. A robust estimator is obtained by replacing $\text{cov}(Y)$ by $\tilde{Y}\tilde{Y}^T$, namely the cross-product matrix of the raw

Table 3.5 Robust (sandwich) standard errors for analysis A in Table 3.4

Parameter	Estimate	Model based s.e.	Robust s.e.
Fixed			
Constant	0.32		
Religion			
Roman Catholic	−0.80	0.176	0.225
Protestant	−0.27	0.098	0.102
Other	−0.63	0.127	0.121
Year			
1984	−0.29	0.048	0.050
1985	−0.06	0.048	0.061
1986	0.06	0.048	0.047
Random			
Level 3: σ_v^2	0.03	0.030	0.020
Level 2: σ_u^2	0.37	0.043	0.039
Level 1: σ_e^2	0.31	0.016	0.022

residuals, which is a consistent estimator of V. This is done for each highest level block of V in order to satisfy the block diagonality structure of the model. This estimator is a generalization of the estimator given by Royall (1986) for a single-level model which uses only the diagonal elements of $\tilde{Y}\tilde{Y}^T$. It is often known as a 'sandwich' estimator from the form of the right-hand side in (3.16).

For the random parameters an analogous result holds. It is also possible to derive robust estimators for residuals, but these generally are not useful because the estimate for each residual corresponding to a higher level unit uses the corresponding value of $\tilde{Y}\tilde{Y}^T$ and this can give very unstable estimates.

We now apply (3.16) to the abortion data analyses and Table 3.5 shows the result for analysis A of Table 3.4 and an OLS analysis. The major change is in the estimate of the standard error for level 1, with only moderate changes for the fixed parameters.

Another approach to providing robust standard errors is to use jackknifing (Miller, 1974). Thus, if we wished to calculate the standard error for a level 2 variance in a model with m level 2 units, the jackknife procedure would involve recomputing the variance for m subsamples, each one formed by omitting one level 2 unit, and using the set of these to form the standard error estimate. The procedure also gives a revised estimate of the parameter itself. Longford (1993; see Chapter 6) gives an example in the analysis of a complex matrix sample design and suggests that there may be often a considerable loss of efficiency using the jackknife method, and it is also computationally intensive. A more flexible method is that of bootstrapping which we now describe.

3.6 The bootstrap

For a general introduction to bootstrapping see Efron and Gong (1983) and Laird and Louis (1987, 1989), and Davison and Hinkley (1997) for more extensive discussions, especially in the context of a multilevel model. Three general bootstrapping approaches are available: a fully non-parametric bootstrap, a semi-parametric or 'residuals' bootstrap and a fully parametric bootstrap.

3.6.1   The fully non-parametric bootstrap

The basic non-parametric bootstrap procedure for a single-level model involves simple random resampling with replacement of the (level 1) units to generate a single bootstrap sample. The model parameter estimates are then re-estimated for this sample. This procedure is repeated a large number (N) of times yielding N sets of parameter estimates which are then treated as a simple random sample and used to derive standard errors or confidence intervals. For a multilevel model, however, such a procedure is inadequate since it assumes identically distributed responses. An alternative that does provide suitable estimates is to resample only the higher level units, but if the number of such units is small this will not be very efficient. Nevertheless where the number of highest level units is large, for example typically in a repeated measures model, this procedure is useful.

No other procedure, for example for a 2-level model sampling at level 2 and then at level 1 within level 2 units, seems able to preserve the original correlation structure since the bootstrap resampling procedure assumes independent responses for the level 1 units, and this is not the case for a multilevel structure. To preserve the correlation structure we should sample the full set of (correlated) level 1 units within each level 2 unit, that is simply to sample the higher level units.

3.6.2   The fully parametric bootstrap

The fully parametric bootstrap utilizes the distributional assumptions of the model in order to generate simulated values which are used to estimate bootstrap sets of parameters. Consider the simple 2-level model assuming Normality

$$y_{ij} = (X\beta)_{ij} + u_j + e_{ij}, \quad u_j \sim N(0, \sigma_u^2), \quad e_{ij} \sim N(0, \sigma_e^2)$$

To generate a bootstrap sample we select at random from $N(0, \sigma_u^2)$ a set of level 2 values u_j^* and for each level 2 unit a set of e_{ij}^* from $N(0, \sigma_e^2)$. These are added to $(X\beta)_{ij}$ to generate a set of pseudo values y_{ij}^* which are then treated as a set of responses from which a new set of bootstrap parameter values, $\hat{\beta}^*, \hat{\sigma}_u^{*2}, \hat{\sigma}_e^{*2}$ is obtained.

Once the set of bootstrap values is available we can use these to estimate the parameter covariance matrices or standard errors using the usual sample procedures. Confidence intervals for the original parameter estimates or functions of them can be constructed from these by assuming Normality. Alternatively we can construct intervals non-parametrically from the percentiles of the set of empirical bootstrap values, and where the median value for a parameter or function of parameters deviates substantially from the original parameter estimate a bias correction procedure should be used. This involves smoothing the bootstrap distribution using, say, a standard Normal distribution. We first estimate z_0 which is the standard Normal score corresponding to the percentile position of the original parameter estimate. Writing $z^{(1-\alpha)}$, $z^{(\alpha)}$ for the standard Normal deviates corresponding to the required (symmetric) percentiles (for example 5% and 95%) we transform back to the bootstrap distribution from the standard Normal distribution values

$$2z_0 + z^{(1-\alpha)}, 2z_0 + z^{(\alpha)}$$

Efron (1988) discusses this and a further correction based on skewness to improve accuracy.

Table 3.6 Bootstrap standard errors and 90% confidence intervals for Analysis A in Table 3.4. Bootstrap uses 1000 simulated datasets

Parameter	Model-based s.e.	Bootstrap s.e.	Normal CI	Non-parametric adjusted CI
Fixed				
Religion				
Roman Catholic	0.176	0.173	(−1.084, −0.516))	(−1.128, −0.532)
Protestant	0.098	0.100	(−0.429, −0.101)	(−0.420, −0.106)
Other	0.127	0.132	(−0.846, −0.414)	(−0.805, −0.377)
Year				
1984	0.048	0.048	(−0.365, −0.209)	(−0.374, −0.216)
1985	0.048	0.047	(−0.140, 0.014)	(−0.141, 0.012)
1986	0.048	0.048	(−0.015, 0.141)	(−0.019, 0.141)
Random				
Level 3: σ_v^2	0.030	0.022	([0], 0.066)	(0, 0.080)
Level 2: σ_u^2	0.043	0.041	(0.302, 0.436)	(0.308, 0.438)
Level 1: σ_e^2	0.016	0.015	(0.284, 0.334)	(0.288, 0.336)

If we wish to obtain bootstrap estimates for estimated level 2 residuals then for each bootstrap sample we also estimate the residuals, $\hat{u}_j^*$. To estimate the 'comparative' variance of the residuals for each level 2 unit we need to work with $\tilde{u}_j^* = \hat{u}_j^* - u_j^*$ and then use these directly to estimate the required variance, or covariance matrix where there are several random coefficients. They can also be used to construct non-parametric confidence intervals as above.

Table 3.6 gives parametric bootstrap estimates of standard errors and a central 90% confidence interval based upon a Normality assumption and also a non-parametric estimation from 1000 bootstrap samples for the model of Table 3.5.

The bootstrap standard errors agree quite well with the model-based ones, except for the level 3 variance. This parameter is based upon only 54 level 3 units as opposed to 264 level 2 and 1056 level 1 units. This is reflected also in the bootstrap confidence intervals where the non-parametric intervals are fairly close to those of the Normal theory except for the level 3 variance. In general, despite the computational overhead, bootstrap intervals will be desirable where effective sample sizes are small, especially for the random parameters. Where distributions are markedly non-Normal the non-parametric intervals are to be preferred, and 1000 is often regarded as a minimum.

3.6.3 The iterated parametric bootstrap and bias correction

In section 2.5 we remarked that, in small samples, RIGLS (REML) estimation may be preferred since it produces unbiased estimates for the random parameters. Table 3.7 shows the comparison of the two estimates with a simple variance components model for the data of Table 3.1.

Since the number of level 2 units is moderately large the bias in the IGLS variance estimates is small, about 3%. We can avoid this bias by carrying out a RIGLS estimation but we can also use the bootstrap as a bias correction tool, and we do this here for illustration only.

Table 3.7   JSP mathematics data. Normal score transform of 11-year score; original 8 year score centred about mean

Parameter	Estimate (s.e.)		
	IGLS	RIGLS	Bootstrap (1000)
Fixed			
Constant	0.137	0.137	0.136
8-year score	0.095 (0.004)	0.095 (0.004)	0.095 (0.004)
Gender (boys–girls)	−0.044 (0.050)	−0.044 (0.050)	−0.044 (0.049)
Social class (non-manual–manual)	−0.153 (0.057)	0.154 (0.057)	−0.152 (0.057)
Random			
Level 2: σ_u^2	0.0808 (0.0234)	0.0833 (0.0239)	0.0780 (0.0230)
Level 1: σ_e^2	0.4234 (0.0229)	0.4253 (0.0230)	0.4213 (0.0224)

Assume we have carried out an IGLS estimation and obtained the results in column 1 of Table 3.6, we have

$$y_{ij} = (X\beta)_{ij} + u_j + e_{ij}$$

$$u \sim N(0, \sigma_u^2), \quad e \sim N(0, \sigma_e^2) \tag{3.17}$$

with estimates for the parameters $\hat{\theta}^T = (\hat{\beta}, \hat{\sigma}_u^2, \hat{\sigma}_e^2)$.

We now carry out a bootstrap and obtain the mean of the bootstrap parameter estimates $\bar{\theta}$, say. If, after a large number of bootstraps, $\bar{\theta}$ is different from $\hat{\theta}$ then the difference is an estimate of the bias of the estimation procedure we have used. In other words, assuming that the original estimates are in fact the 'true' model parameters, simulating from these and estimating parameters on average gives different parameter estimates from the original, implying that our estimation procedure is biased. The estimate of the bias is simply $\bar{\theta} - \hat{\theta}$, and a bias-corrected estimate is $2\hat{\theta} - \bar{\theta}$. We see from the third column of Table 3.7 that, using a bootstrap sample of 1000 we estimate a downward bias for the level 2 variance of $0.0808 - 0.0780 = 0.0028$ and similarly for the level 1 variance a downward bias of 0.0021. The bias corrected estimates are therefore 0.0836 and 0.4255 respectively, which are, as expected, close to the RIGLS estimates. If we wish to calculate quantiles etc. from the set of bootstrap replicates then we should apply the bias correction to these estimates also.

In many cases the bias associated with an estimation procedure is a function of the true parameters. In this case, since we are simulating from the estimated parameters our bias estimate will itself generally be biased. In this case we can adopt an 'iterative bias correction'. This works as follows.

Carry out a bootstrap replicate set and obtain a bias-corrected vector of parameters, say $\hat{\theta}_1$. This will still, generally, be biased but less so than the original. Carry out a second bootstrap replicate set using the parameters $\hat{\theta}_1$ instead of the original estimates and calculate a new bias-corrected estimate $\hat{\theta}_2$. While, generally, still biased, the bias will tend to be less than the previous bias. Repeat this using the new bias-corrected estimates until the sequence of estimates converges. In some cases convergence may not occur, but if it does we will obtain unbiased estimates (Kuk, 1995). This procedure is most useful when applied with discrete response models based upon quasilikelihood estimation, where in certain circumstances large biases are produced. A detailed description of the procedure for such models is given by Goldstein and Rasbash (1996) and we will use this in an example in Chapter 4.

3.6.4   The residuals bootstrap

One potential drawback to the fully parametric bootstrap is that it relies upon the (Normal) distribution assumption for the residuals. In single level linear models a residuals (semi-parametric) bootstrap can be implemented by fitting the model, calculating the empirical residuals by subtracting the predicted response from the observed and then for each bootstrap iteration sampling from these residuals with replacement. In a multilevel model, however, the situation is more complicated. Consider first a 'crude' residuals bootstrap as follows for the model (3.17):

1. Estimate residuals $(\hat{u}_j, \hat{e}_{ij})$.
2. Sample with replacement m level 2 residuals and N level 1 residuals, add these to the fixed part estimate to generate a new set of Y values.
3. Fit the model to these new data and obtain the parameter estimates.
4. Repeat steps 2 and 3, say 1000 times.

Such a procedure leads to biases because the residuals are shrunken and the estimates across levels are correlated, negatively in the present case, so independent resampling is inappropriate. We require, therefore, both to 'reflate' residuals and create independence before resampling.

Consider first just reflating the residuals separately at each level. We illustrate with the 2-level model

$$y_{ij} = (X\beta)_{ij} + (ZU)_j + e_{ij}$$
$$U^T = \{U_0, U_1 \ldots\} \tag{3.18}$$

Having fitted the model we estimate the residuals for each level 2 unit j

$$\{\hat{u}_{0j}, \hat{u}_{1j} \ldots\}, \hat{e}$$

For convenience we shall illustrate the procedure using the level 2 residuals, but analogous operations can be carried out at all levels. Write the empirical covariance matrix of the estimated residuals at level 2 in (3.18) as

$$S = \frac{\hat{U}\hat{U}^T}{m} \tag{3.19}$$

and denote the corresponding model estimated covariance matrix of the random coefficients at level 2 as R. The empirical covariance matrix is estimated using the number of level 2 units, m, as divisor rather than $m - 1$. We assume that the estimated residuals have been centred, although centring will only affect the overall intercept value. We also note that no account is taken of the relative sizes of the level 2 units and we could consider a weighted form of (3.19)

$$S = \frac{\hat{U}W\hat{U}^T}{N} \tag{3.20}$$

where W is the $(m \times m)$ diagonal matrix with unit sizes on the diagonal. This is equivalent to defining new residual variables

$$U'_{hj} = U_{hj}\sqrt{mn_j/N} \tag{3.21}$$

and using these with (3.19).

For (3.19) we now seek a transformation of the residuals of the form

$$\hat{U}^* = \hat{U}A$$

where A is an upper triangular matrix of order equal to the number of random coefficients at level 2, and such that

$$\hat{U}^{*^T}\hat{U}^*/m = A\hat{U}^T\hat{U}A = A^T S A = R \tag{3.22}$$

so that these new residuals have the required model covariance matrix, and we now can sample sets of residuals with replacement from $\hat{U}^*$. This will be done at every level of the model, with sampling being independent across levels, thus retaining the independence assumption of the model. Having sampled a set of these residuals we back-transform these using $\hat{U} = \hat{U}^* A^{-1}$ and add to the fixed part of the model, along with the corresponding level 1 residuals to obtain the new set of responses.

We can form A as follows: write the Cholesky decomposition of S, in terms of a lower triangular matrix as $S = L_S L_S^T$ and the Cholesky decomposition of R as $R = L_R L_R^T$. We have

$$L_R L_S^{-1}\hat{U}^T\hat{U}(L_R L_S^{-1})^T = L_R L_S^{-1}S(L_S^{-1})^T(L_R)^T = L_R(L_R)^T = R$$

and the required matrix is therefore $A = (L_R L_S^{-1})^T$.

Carpenter et al. (1999) use this procedure, unweighted, to demonstrate the improved (confidence interval based) coverage probabilities compared to the parametric bootstrap when the level 1 residuals have a chi-squared distribution rather than a Normal. Further work confirms this but also suggests that the procedure may underestimate coverage for certain departures from an assumed Normal distribution. (Carpenter et al., 2002). If we apply this procedure to the data in Table 3.6 we obtain estimates very similar to those using the fully parametric bootstrap. This procedure can also be iterated in the same way as the parametric bootstrap.

The above reflating procedure takes no account of dependencies across levels. If we now write

$$Q^T = \{U_0, U_1 \dots, e\}, \quad Q^T \text{ is } (N \times [p+1]) \tag{3.23}$$

where p is the number of level 2 random effects and Q^T is the length of the data matrix, then analogously to (3.19) we form

$$S = \frac{\hat{Q}\hat{Q}^T}{N} \tag{3.24}$$

and proceed as before with computing transformed residual sets for the resampling bootstrap. The R matrix has the form

$$\begin{pmatrix} \Omega_u & 0 \\ 0 & \sigma_e^2 \end{pmatrix}$$

so that the Cholesky decomposition is formed in the same way as the separate ones above. The S matrix, however, does not have this form since there are cross-product terms for the level 2 and level 1 residuals of the form

$$\frac{1}{N} \sum_j \hat{u}_{hj} \sum_i \hat{e}_{ij} \tag{3.25}$$

We note that the Q vectors are still uncorrelated across levels so that we can sample them separately at each level and maintain the model independence assumption as before. We also note that if we ignore the term (3.25) we obtain the weighted procedure given by (3.20). Research is currently being carried out using these extensions that additionally take into account the cross-level correlations.

3.7 Aggregate level analyses

As we discussed in Chapter 1, there are sometimes occasions when the only data available for analysis have already been aggregated to a higher level. For example, we may have information on student achievement in terms only of the mean achievement for each school, or information on utilization of health services in terms only of the total number of episodes for each administrative area. We now examine the possibilities for carrying out analyses with aggregate-level data and explore how far these can provide information about the parameters of a more disaggregated model.

Consider the simple model for the Junior School Project data with the 11-year mathematics test score as response and the earlier mathematics score as a covariate

$$y_{ij} = \beta_0 + \beta_1 x_{ij} + u_j + e_{ij} \tag{3.26}$$

Suppose that we now aggregate to the school level by averaging over all pupils in each school to obtain

$$y_{.j} = \beta_0 + \beta_1 x_{.j} + u_j + e_{.j} \tag{3.27}$$

Fitting this as a single-level model, the level 1 variance is $\sigma_u^2 + n_j^{-1}\sigma_e^2$ and we can fit the model by specifying two explanatory variables for the random part, namely

$$z_0 = 1, \quad z_{1j} = n_j^{-0.5}$$

with random coefficients e_{0j}, e_{1j} each having a variance with zero covariance. In many surveys the same number of level 1 units will be sampled from each level 2 unit, in which case a single explanatory variable z_0 will suffice, but this does not then allow us to estimate the level 1 and level 2 variances separately. The other problem with such an analysis is that the estimates will be inefficient compared with those from a 2-level model based on individual student data.

Analysis A in Table 3.8 gives the results of an analysis using just the single explanatory variable z_0 and analysis B additionally uses z_{1j} and so is equivalent to a single-level weighted regression model. In both analyses we have included the proportion of non-manual students and the proportion of girls as explanatory variables, that is the average values of the corresponding (0.1) dummy variables.

Table 3.8   School level analysis of JSP data

Parameter	Estimate (s.e.)		
	A	B	C
Fixed			
Constant	0.18	0.16	0.16
8-year score	0.091 (0.019)	0.092 (0.020)	0.094 (0.021)
Gender (Propn. boys)	−0.34 (0.30)	−0.31 (0.30)	−0.29 (0.29)
S. Class (Propn. N.M.)	0.00 (0.20)	0.00 (0.28)	−0.01 (0.27)
Random			
σ^2_{u0}	0.11 (0.021)	0.11 (0.040)	0.08 (0.024)
σ^2_{e0}		0.08 (0.37)	−
σ_{u01}			0.00 (0.01)
σ^2_{u1}			0.004 (0.004)
−2 (log-likelihood)	31.33	31.28	29.44

In comparison with the analyses in Table 3.7 while the coefficient of the 8-year maths score changes only slightly, those for gender and social class change markedly. We also see how the standard errors are substantially greater. In fact, although the number of students per school varies between 3 and 49, the inclusion of z_{1j} has little effect.

For these data we know that the slope of the 8-year score is random across schools. In this case model (3.27) becomes

$$y_{\cdot j} = \beta_0 + \beta_1 x_{\cdot j} + u_{0j} + u_{1j} x_{\cdot j} + e_{\cdot j} \qquad (3.28)$$

and we obtain the additional contributions to the variance of the aggregated level 2 units

$$\sigma^2_{u1} x^2_{\cdot j}, \quad 2\sigma_{u01} x_{\cdot j}$$

Analysis C in Table 3.8 shows the results of fitting this model, giving a poor estimate of the random coefficient variance, and unlike analysis B it is not possible to estimate a separate level 1 variance because of the small number of units in the analysis.

If there is complex variation at level 1, such as we fitted in Table 3.2, then for such an explanatory variable, say z_{2ij}, we would obtain the further contributions to the variance for the aggregated model for unit j

$$2\sigma_{e02} z_{2 \cdot j} / n_j, \quad \sigma^2_{e2} \sum_i z^2_{2ij} / n^2_j$$

The first of these terms can be fitted as a covariance and the second as a variance, by defining appropriate explanatory variables. In the present case the data are not extensive enough to allow us to fit these additional variables. We also note that the values of the squared explanatory variables in the second of these expressions will often not be available for aggregated data.

If we have an initial 3-level model, and data are aggregated to level 2, we need to specify properly the level 2 random variation resulting from the aggregation process. Failure to do this, may allow us to fit random variation at level 3, but any interpretation of this may be problematical because it may have arisen solely as a result of misspecifying

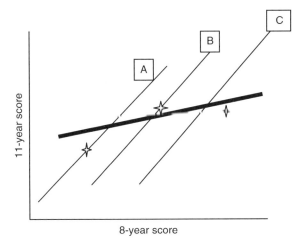

Figure 3.7 Aggregate vs. individual-level model inferences; hypothetical illustration for 3 schools.

the variation at level 2. For example, if we have an explanatory variable which is strongly correlated with the size of the level 2 units, and we fail to include a random coefficient for z_{1j} at level 2, we may well be able to fit a random coefficient for it at level 3, but the usual interpretation of such a coefficient would be incorrect.

We now look at what happens to the fixed part coefficients when aggregation takes place and we have already seen that the values of the coefficients for gender and social class change. Consider the 2-level model

$$y_{ij} = \beta_0 + \beta_1 x_{ij} + \beta_2 x_{.j} + u_j + e_{ij} \tag{3.29}$$

where the coefficient for $x_{.j}$ in the aggregated model is now $\beta_1 + \beta_2$. We saw in Table 3.1 that the coefficient for the school mean 8-year score was very small, so that we would expect the coefficient for this in the aggregated model to be similar, which Table 3.8 confirms. For gender and social class the coefficients of the corresponding aggregated variables from a 2-level analysis are respectively -0.06 and -0.09, which when added to the (non-aggregated) coefficients for gender and social class give values of -0.09 and -0.06 respectively. These are rather different from those in Table 3.6, but the standard errors are very large. Where there is a contextual or compositional effect, whether through the mean aggregated value, or some other statistic derived from the student level distribution, then an aggregated analysis will not allow us to obtain separate estimates for the individual and compositional coefficients.

3.7.1 Inferences about residuals from aggregate level analyses

The use of aggregate level analysis can be particularly misleading for residual estimates. Woodhouse and Goldstein (1989) show how this can occur for educational data aggregated to the level of Education Authority and Figure 3.7 illustrates a situation where the use of aggregate data at face value provides incorrect inferences. For the aggregate analysis the school ordering for given 8-year score is: $A < B < C$. For the pupil-level analysis the school ordering is: $A > B > C$, which gives a quite different conclusion. In general, the only circumstance where we can be certain that aggregate- and individual-level inferences are the same is where the explanatory variable distribution (that of the 8-year score here) is the same for each higher level unit (school).

In summary, we have seen that it is sometimes possible to model aggregated data, but this has to be carried out with care, and any interpretations will be constrained by the nature of the true, underlying, non-aggregated model. In addition, the precisions of the estimates obtained from an aggregated analysis will generally be much lower than those obtained from a full multilevel analysis. A discussion of the aggregation issue can also be found in Aitkin and Longford (1986).

3.8   Meta analysis

The term Meta Analysis (Hedges and Olkin, 1985) refers to the pooling of results of separate studies, all of which are concerned with the same research hypothesis. The aim is to achieve greater accuracy than that obtainable from a single study and also to allow the investigation of factors responsible for between-study variation. Each study typically provides an estimate for an 'effect', for example a group difference, for a 'common' response and the original data are often unavailable for analysis. In general the response measures used will vary, and care is needed in interpreting them as meaning the same thing. Furthermore, the scales of measurement may differ, so that the effect is usually standardized using a suitable within-study estimate of between-unit standard deviation. If the study result derives from a multilevel model, then this estimate will be based on the level 1 variance, or where this is complex on an estimate pooled over the effect groups being compared. It is important that comparable estimates are used from each study. This implies that the specification of the level 1 units is comparable and that the sources of higher level variation are properly identified. For example for a set of studies comparing teaching methods using a number of schools, the within-school between-student variation would be appropriate for standardization, which implies ideally that the studies concerned should provide estimates of this using suitable multilevel techniques. We consider the case where only a single effect is of interest, but the generalization to the multivariate case is straightforward (see Chapter 6).

In the standard case, for the j-th study we define the standardized effect d_j where this is a dimensionless quantity. It may, for example, be a correlation coefficient, a standardized regression coefficient, a group difference or a weighted group difference. We require an estimate of the variance of d_j, say σ_j^2, and more generally we require the variance of a dimensionless function having the general form

$$\sum_h w_{hj}\hat{\beta}_{hj}/\hat{\sigma}_{ej} \tag{3.30}$$

where the $\hat{\beta}_{hj}$ are parameter estimates from the j-th study. A pooled estimate can now be derived using these standardized effects and their variance estimates. For moderately large numbers of level 1 units, we can ignore the variation in the estimate of the level 1 standard deviation ($\hat{\sigma}_{ej}$) and calculate the variance of the numerator of (3.30) using the estimated covariance matrix of the $\hat{\beta}_{hj}$.

In the general case we may have several studies each of which may provide information on treatment effects at different levels of aggregation. We can formulate a general class of meta analysis models by considering a simple 2-level structure. Assume that we have a collection of studies, each concerned with the comparison of several 'treatments'. These treatments may be distinct categories (represented by dummy variables) or may be effects represented by regression coefficients or a mixture of the two kinds. The basic models we shall develop are 'variance component' models but we will also illustrate a random coefficient model, and the variance heterogeneity (complex variation) case can also be incorporated.

For the i-th subject in the j-th study who received the h-th treatment, we can write a basic underlying model for outcome y_{hij} as

$$y_{hij} = (X\beta)_{ij} + \alpha_h t_{hij} + u_{hj} + e_{hij}$$
$$h = 1, \ldots, H; \quad j = 1, \ldots, J; \quad i = 1, \ldots, n_{hj} \qquad (3.31)$$
$$u_{hj} \sim N(0, \sigma_{hu}^2); \quad e_{hij} \sim N(0, \sigma_{he}^2)$$

where $(X\beta)_{ij}$ is a linear function of covariates for the i-th subject in the j-th study, u_{hj} is the random effect of the h-th treatment for the j-th study and e_{hij} is the random residual of the h-th treatment for subject i in study j. The term t_{hij} is a dummy treatment variable (contrasted against a suitable base category) and α_h is the treatment contrast of primary interest. If the treatment dummy variables are replaced by a continuous variable t_{ij} then (3.31) becomes

$$y_{ij} = (X\beta)_{ij} + \alpha t_{ij} + u_j + e_{ij}$$
$$j = 1, \ldots, J; \quad i = 1, \ldots, n_j$$
$$u_j \sim N(0, \sigma_u^2); \quad e_{ij} \sim N(0, \sigma_e^2)$$

It is also clearly possible to allow the variances within and between studies to be different for each treatment or to vary with the value of a continuous treatment variable, leading to complex variance structures. We can also introduce covariates where data are available and appropriate, and interactions between treatments and covariates. For example, a particular treatment contrast may differ according to covariate values. We may also relax the Normality assumption of the level 1 residuals, for example if fitting a generalized linear multilevel model (for an example see Turner et al., 2000).

3.8.1 Aggregate and mixed level analysis

Consider now the case where (3.31) is the underlying model but we only have data by treatment group at the study level. Aggregating to this level, as in section 3.7, we write the mean response as

$$y_{h\cdot j} = (X\beta)_{\cdot j} + \alpha_h t_{h\cdot j} + u_{hj} + e_{h\cdot j} \qquad (3.32)$$

where the '.' notation denotes the mean for study j. A difficulty may arise with the first term in (3.32) since this implies that the mean of the covariate function $(X\beta)_{ij}$ for each study is available.

The corresponding model for the case of a continuous treatment variable is

$$y_{\cdot j} = (X\beta)_{\cdot j} + \alpha t_{\cdot j} + u_j + e_{\cdot j}$$

Consider the special case of two treatments, $h = 1, 2$. We collapse (3.32) and, using an obvious notation, rewrite to give

$$y'._j = y_{1 \cdot j} - y_{2 \cdot j} = \alpha + u'_j + e'._j$$
$$\alpha = \alpha_1 - \alpha_2$$

(3.33)

This implies $\text{var}(u'_j) = \text{var}(u_{1j}) + \text{var}(u_{2j}) - 2 \text{cov}(u_{1j}, u_{2j})$. We can also combine (3.31) and (3.32) into a single model for the case where some aggregated responses are in terms of separate treatment groups and some are in terms of contrasts of groups.

3.8.2 Defining origin and scale

When combining data from aggregate-level studies it is necessary to ensure that the response variable scales are the same and that there is a common origin. In traditional two-treatment meta analyses the treatment difference is divided by a suitable (pooled) within-treatment standard deviation. In our general model, likewise, the response variable in each study can be scaled by dividing it by an estimate of the level 1 standard deviation. Where individual data are available we may use an estimate of the level 1 standard deviation from a preliminary analysis and for aggregate data we may derive this from reported summary information, if this is available.

In situations where the same response variable is used in each study, and scaling has been carried out, we can apply (3.32) and (3.33) directly. In many cases, however, different response variables are used. For example, in class size studies different reading tests may be used. In this case we would not generally expect the means for corresponding treatments to be identical. One procedure for dealing with this is to choose one treatment as a reference treatment (or control) and in each study subtract its mean from the values of the other treatments and work *with these differences*. This is the standard approach in two-treatment studies and such differences can either be fixed effects or vary randomly across studies.

3.8.3 An example: meta analysis of class size data

Goldstein et al. (2000b) consider the meta analysis of nine studies of the effects of class size on the learning of young children. Eight of these studies report only aggregate-level data and one (STAR) provides individual student data. The studies were selected from thousands of reported studies, as being those that satisfied stringent quality criteria for design and reporting of data. Applying the above models, after suitable scaling, a joint analysis gives the results shown in Table 3.9.

In the analysis using only the STAR individual-level data we can distinguish three levels: individual, class and school. The response has been normalized so that the level 1 variance is fixed at 1.0 and we see that there is an overall negative relationship with class size whereby there is a gain of about a quarter of a standard deviation for a reduction in class size of 10 children. There is also reasonably large variation between classes, within schools and between schools in the magnitude of this effect. The STAR study was a randomized controlled trial and its design raises some particular issues of interpretation which are discussed by Goldstein and Blatchford (1998).

In the combined analysis the eight non-randomized, but longitudinal, studies reporting at aggregate level are combined with the STAR study. This allows us to estimate between-study variation at level 4 and there is some evidence for the relationship

Table 3.9 Combined analysis of class size study data. Reading score response at end of year 1 (post kindergarten)

	STAR data only	*Combined data*
Fixed		
α_0	0.078	0.184
α_1 (class size, linear)	−0.024 (0.006)	−0.022 (0.007)
α_2 (pre-test)	0.907 (0.018)	0.907 (0.018)
Random		
Level 4 (between study)		
σ^2_{w0}		0.038 (0.020)
σ_{w01}		−0.004 (0.002)
σ^2_{w1}		0.0004 (0.0002)
Level 3 (between school)		
σ^2_{v0}	0.305 (0.064)	0.305 (0.064)
σ_{v01}	0.00014 (0.004)	0.00012 (0.004)
σ^2_{v1}	0.0006 (0.0006)	0.0006 (0.0006)
Level 2 (between class)		
σ^2_{u0}	0.139 (0.023)	0.138 (0.023)
Level 1 (between student)		
σ^2_e	1.000 (0.023)	1.000 (0.023)
−2 (log-likelihood)	11996.5	11948.3

varying across studies, although the combined mean estimate differs little from that for the STAR study alone.

3.8.4 Practical issues

There are a number of practical problems with meta analysis studies. One of these is where the sample of studies used is subject to systematic bias. This can occur, for example, if some studies do not provide sufficient data to estimate a standardized difference and they are a special group. Another common problem arises where the analysis is based upon published studies and those studies which found 'non-statistically significant' results tend to remain unpublished. This implies that the distribution of results is censored with the smaller ones tending to be missing, a situation known as the publication bias effect. Vevea (1994) discusses the possibility of weighting the studies, that is the units in the model (3.14), using a suitable function of the statistical significance level associated with each effect, in order to compensate for the selective exclusion. Thus we could carry out a weighted analysis (see section 3.4) where the weights are, say, proportional to the significance level. Vevea also considers the possibility of estimating the weights. We can also consider weighting of the (study) level 2 units, and extensions to differential weighting of level 1 units are possible. These weights may reflect information about study quality or possibly non-response. Such an analysis might be undertaken as a sensitivity analysis to complement an equally weighted analysis. The weighting procedure described in section 3.4 can be used.

3.9 Design issues

When designing a study where the multilevel nested structure of a population is to be modelled, the allocation of level 1 units among level 2 units and the allocation of these among level 3 units etc. will clearly affect the precision of the resulting estimates of both the fixed and random parameters. The situation becomes more complex, for example, when there are cross-classifications and where there are several random coefficients. There are generally differential costs associated with sampling more level 1 units within an existing level 2 unit as opposed to selecting further level 1 units in a new level 2 unit.

Some approximations for studying the standard errors of the fixed coefficients have been derived by Snijders and Bosker (1993) in the case of a simple 2-level variance components model. They are concerned with students sampled within schools and assume that the cost of selecting a student in a new school is a fixed constant times the cost of selecting a student in an already selected school. They also assume that there is a minimum of 11 students per school. They tend to find that, where this constant is greater than 1 and the total number of students to be sampled is fixed, the sample of schools should be as large as possible, although this will not necessarily be true for all the coefficients of interest.

Moerbeek et al. (2000) consider a 3-level model where interest focuses on the estimation of a treatment effect so that the aim is to minimize the variance of the relevant fixed coefficient. They consider the balanced case and show analytically how differential allocations across units can minimize overall cost. In particular they consider the allocation of different treatments to level 1 units within level 2 or level 3 units versus allocation at the higher unit level. Moerbeek et al. (2001) extend their analysis to the 2-level binary response model (see Chapter 4) and show how similar results can be obtained using MQL estimation. They show, in the case of a particular simulation study, that design decisions based upon MQL are similar to those when either PQL or maximum likelihood estimation is used. They also discuss practical and ethical issues associated with allocation within higher level units, especially where these are for example schools, where 'contamination' and lack of independence can arise (see also Goldstein and Blatchford (1998) for a discussion in the context of educational interventions).

In practice balanced designs are a special case and interest focuses on both fixed and random parameters. Analytical results are then generally not available but a guide to design efficiency can be obtained by simulating the effect of different design strategies and studying the resulting characteristics of the parameter estimates, such as their mean squared errors. This will be time consuming however, since for each design a number of simulated samples will be required. Mok (1995) carries out such an analysis for a 2-level random coefficient model and makes tentative recommendations based upon the intra-unit correlation and numbers of level 1 and level 2 units.

Cohen (1998) considers the optimality (efficiency) with respect to various fixed and random effects in a 2-level balanced model as a function of the intra-school correlation and shows that optimality criteria can vary considerably with the choice of parameter. Moerbeek and Wong (2002) extend these results by considering the robustness to misspecification of the intra-unit correlation and also consider ways of combining optimal choices across parameters. They consider the case where weights are chosen to minimize a linear function of parameter variances. It is also possible to consider an optimality criterion based, for example, upon the determinant of the covariance matrix of the fixed or random parameters.

4

Multilevel Models for Discrete Response Data

4.1 Generalized linear models

All the models of previous chapters have assumed that the response variable is continuously distributed. We now look at data where the response is essentially a count of events. This count may be the number of times an event occurs out of a fixed number of 'trials', in which case we usually deal with the resulting proportion as response: an example is the proportion of deaths in a population, classified by age. We may also have a vector of counts representing the numbers of events of different kinds which occur out of a total number of events: an example is given in Chapter 3 where we studied the number of responses to each, ordered, category of a question on abortion attitudes.

Statistical models for such data are referred to as 'generalized linear models' (McCullagh and Nelder, 1989). A 2-level model can be written in the general form

$$\pi_{ij} = f(X\beta)_{ij} \qquad (4.1)$$

where π_{ij} is the expected value of the response for the ij-th level 1 unit and f is a nonlinear function of the 'linear predictor' $X_{ij}\beta_j$. Note that we allow random coefficients at level 2. The model is completed by specifying a distribution for the *observed* response $y_{ij}|\pi_{ij}$. Where the response is a proportion this is typically taken to be binomial and where the response is a count taken to be Poisson. It remains for us to specify the nonlinear 'link' function f. Table 4.1 lists some of the standard choices, with logarithms chosen to base e. We shall later see how functions of these, applied for example to ordered categories, can be specified.

In addition to these we can also have the 'identity' function $f^{-1}(\pi) = \pi$, but this can create difficulties since it allows, in principle, predicted counts or proportions which are respectively less than zero or outside the range $(0, 1)$. Nevertheless, in many cases, using the identity function produces acceptable results which may differ little from those obtained with the nonlinear functions. In the following sections we consider each common type of model in turn with examples. We shall also be introducing new estimation procedures based upon maximum likelihood, MCMC, and quasilikelihood approximations.

Table 4.1   Some basic nonlinear link functions

Response		Name
Proportion	$f^{-1} = \log\{(\pi)/(1-\pi)\}$	Logit
Proportion	$f^{-1} = \log\{-\log(1-\pi)\}$	Complementary log-log
Proportion	$\pi = \int\limits_{-\infty}^{(X\beta)} \phi(t)dt$	Probit
Vector of proportions	$f^{-1} = \log(\pi_s/\pi_t) \ (s = 1, \ldots, t-1)$	Multivariate logit
Vector of proportions	$f^{-1} = \log\left\{\log\left(\dfrac{1}{t-1} + \dfrac{\pi^{(s)}}{\pi^{(t)}}\right)\right\}$	Multivariate complementary log-log
	$= (X\beta), \ (s = 1, \ldots, t-1)$	
	$\pi^{(s)} = \left(e^{e^{(X\beta)^{(s)}}} - \dfrac{1}{t-1}\right)\left(\sum\limits_{h=1}^{t-1} e^{e^{(X\beta)^{(h)}}}\right)^{-1}$	
Count	$f^{-1} = \log(\pi)$	Log

4.2   Proportions as responses

Consider the 2-level variance components model with a single explanatory variable where the expected proportion is modelled using a logit link function

$$\pi_{ij} = \{1 + \exp(-[\beta_0 + \beta_1 x_{1ij} + u_{0j}])\}^{-1} \tag{4.2}$$

The observed responses y_{ij} are proportions with the standard assumption that they are binomially distributed

$$y_{ij} \sim Bin(n_{ij}, \pi_{ij}) \tag{4.3}$$

where n_{ij} is the denominator for the proportion. We also have

$$\text{var}(y_{ij}|\pi_{ij}) = \pi_{ij}(1 - \pi_{ij})/n_{ij} \tag{4.4}$$

To emphasize the link with continuous responses and to introduce quasilikelihood estimation we now write the level 1 component as

$$y_{ij} = \pi_{ij} + e_{ij}z_{ij}, \quad z_{ij} = \sqrt{\pi_{ij}(1 - \pi_{ij})/n_{ij}}, \quad \sigma_e^2 = 1 \tag{4.5}$$

Using this explanatory variable Z and constraining the level 1 variance associated with its random coefficient (e_{ij}) to be one, we obtain the required binomial variance in (4.4). When fitting a model we can also allow the level 1 variance to be estimated and by comparing the estimated variance with the value 1.0 obtain a test for 'extra binomial' variation. Such variation may arise in a number of ways.

If we have omitted a level in the model, for example ignored household clustering in a survey with one or more individuals sampled from a household, we would expect a greater than binomial variation at the individual level. Likewise, suppose the individuals and households were nested within areas and we chose to classify individuals, say by gender and three social class groups giving six cells in each area. If we treat these as the level 1 units so that the response is a proportion, then we no longer have a binomial variance since these proportions are based upon the sum of separate binomial variables with differing probabilities. Here the variance for cell j within an area would have the form

$$[\pi_j(1 - \pi_j) - \sigma_1^2]/n_j$$

where n_j is the cell size. To fit such a model we would specify an extra level 1 explanatory variable equal to $1/\sqrt{n_j}$ for the j-th cell, with 'variance' parameter at level 1 which was allowed to be negative (see section 3.1). More generally, we can fit a model with an extra binomial parameter together with a further term such as above to give the following level 1 variance structure (omitting subscripts)

$$[\sigma_0^2 \pi(1 - \pi) + \sigma_1^2]/n$$

We do not, of course, know the true value π_{ij} or π_j so that at each iteration we use estimates based upon the current values of the parameters. Because we are using only the mean and variance of the binomial distribution to carry out the estimation, the estimation is known as 'quasilikelihood' (see Appendix 4.1 for details).

Another way of modelling such extra binomial variation, which has certain advantages, is to insert a 'pseudo level' above level 1. Thus, for individuals sampled within households, level 1 would be that of the individual and we would specify level 2 as that of the individuals also to give exactly one level 1 unit per level 2 unit. We specify binomial variation at level 1 and at level 2 we can now fit further random coefficients. For example, if we fit a random coefficient for the explanatory variable (with a 'variance' parameter which can be allowed to be negative) this is equivalent to specifying an extra level 1 variable $1/\sqrt{n_j}$ as above. In the above example where individuals are classified by gender and social class we can create a level 2 unit coinciding with each level 1 unit, fit binomial variation at level 1 and add level 2 variation which is a function of gender and social class, for example an additive function with four parameters (see Chapter 3). We may wish to model the between-area variation of the cell proportions in terms of a simple variance term, rather as inversely proportional to n_j. In this case we would choose a simple dummy variable structure rather than explanatory variables proportional to $1/\sqrt{n_j}$. This 'pseudo level' procedure is rather similar to the way in which a meta analysis with known level 1 variation is modelled (Chapter 3).

In Appendix 4.1 we describe a linearization procedure that allows us to use existing IGLS and RIGLS estimation for generalized linear models. For the model given by (4.2) and (4.5), a second order Taylor series expansion about the current values of the estimated residuals gives

$$\pi_{ij} = f_{ij}(H_t) - X_{2ij}\beta_{2,t} f'_{ij}(H_t) + (Z_{2ij}\hat{u}_{2j})f'_{ij}(H_t) + X_{2ij}\beta_{2,t+1} f'_{ij}(H_t)$$
$$+ Z_{2ij} u_{2j} f'_{ij}(H_t) + Z_{2ij}(u_{2j} - \hat{u}_{2j})^2 f''_{ij}(H_t)/2$$
$$H_t = X_{2ij}\beta_{2,t} + Z_{2ij}\hat{u}_{2j} \tag{4.6}$$

where the terms on the right-hand side of the first line of (4.6) form an offset subtracted from the response at each iteration. We refer to estimates based upon (4.6) as PQL2 estimates, i.e. predictive quasilikelihood with a second order approximation. If a first order approximation is used these are PQL1 estimates and if the expansion is about zero rather than the current residual estimates $(\hat{u}_{2j})$ these are known as marginal quasilikelihood estimates, MQL2 and MQL1 respectively.

In many applications the MQL procedure will tend to underestimate the values of both the fixed and random parameters, especially where n_{ij} is small. In addition greater accuracy is obtainable if the second order approximation is used rather than the first order based upon the first term in the Taylor expansion. Also, when the sample size is small the unbiased (RIGLS, REML) procedure should be used. To illustrate the difference Table 4.2 presents the results of simulating the following model where the response is binary (0,1). The example assumes one moderate and one fairly large level 2 variance.

$$\text{logit}(\pi_{ij}) = \beta_0 + \beta_1 x_{ij} + u_{0j}$$

$$y_{ij} \sim Bin(1, \pi_{ij})$$

$$\text{var}(u_{0j}) = 0.5, 1.0$$

$$\beta_0 = 0.5, \quad \beta_1 = 1.0$$

There are 50 level 2 units with 20 level 1 units in each level 2 unit. The results in Table 4.2 are based upon 400 simulations of the above model for each variance value.

It is clear that the MQL first order model underestimates all the parameter values, whereas the second order PQL model produces estimates closer to the true values. The estimates given are based upon IGLS. In every case convergence was achieved in less than 10 iterations. Very similar estimates for the fixed coefficients are obtained using RIGLS, and for the level 2 variances the PQL estimates become 0.498 and 0.996 respectively, which are closer to the true values. In addition, the averages of the standard errors given by both models are reasonably close to those calculated empirically from the replications. If we calculate 95% confidence intervals for the parameters in the second order PQL model using the estimated standard errors and assuming Normality then for the variance, we find that about 91% of the intervals include the true value and for β_0 and for β_1 about 95% do so. Hence, inferences about the true values would not be too misleading. The results of Table 4.2 are based upon a balanced dataset with equal numbers of level 1 units within each level 2 unit. In other cases none of these procedures works satisfactorily. Thus, for example, when the average observed probability is very small (or very large), if many of the level 2 units have few level 1 units and there are very few level 2 units with large numbers of level 1 units, we will often find that where the response is binary, there will be many level 2 units where the responses are all zero. In such a case convergence with PQL2 or PQL1 often may not be possible and even where estimates are obtained, they may be substantially biased.

Table 4.2 Mean values of 400 simulations. Empirical standard error in first parentheses; mean of estimated standard errors in second parentheses (IGLS)

Parameter	True $\sigma_{u0}^2 = 0.5$		True $\sigma_{u0}^2 = 1.0$	
	MQL first order	PQL second order	MQL first order	PQL second order
σ_{u0}^2	0.386(0.115)(0.130)	0.480(0.157)(0.152)	0.672(0.157)(0.188)	0.964(0.278)(0.255)
β_0	0.448(0.126)(0.129)	0.499(0.139)(0.138)	0.420(0.145)(0.149)	0.500(0.171)(0.172)
β_1	0.934(0.154)(0.147)	1.018(0.168)(0.154)	0.875(0.147)(0.145)	1.017(0.171)(0.158)

Rodriguez and Goldman (2001) discuss this issue and make recommendations. Where these quasilikelihood methods cannot be used we may carry out a full maximum likelihood estimation (see Appendix 4.2), use MCMC (see Appendix 4.3) or carry out an iterated bootstrap (see Appendix 4.4). All these procedures will be illustrated in the following example.

4.3 An example from a fertility survey

This dataset comes from the 1988 Bangladesh Fertility Survey (Huq and Cleland, 1990) (Table 4.3). It consists of 1934 women who are grouped in 60 districts and the response of interest is whether these women were using contraceptives at the time of the survey. Explanatory variables include age, the number of existing children and whether the district is urban or rural.

The MQL1 estimates are fairly close to the PQL2 estimates and in fact there are only two level 2 units (districts) with less than 10 women. There is a clear relationship with age, younger women being more likely to use contraceptives. Urban women are far more likely to use contraceptives; the odds ratio being $e^{0.82} = 2.3$. In terms of number of children the greatest difference is in moving from no children to one child, an odds ratio of 3.1, with little change after 2 children. There is a relatively large variation in the between-district urban–rural difference. If we allow extra-binomial variation for the PQL2 model the estimate of the multiplicative term is 0.96 with a standard error of 0.03 and if we fit an additive term we obtain a value of -0.09 with a standard error of 0.15. Neither of these estimates provides much evidence for extra binomial variation.

Figure 4.1 shows the standardized residual plots for the intercept and coefficient of urban locality.

These plots produce approximately straight lines. Some care is needed with such plots when the number of level 1 units per higher level unit is small. The residual estimate is a linear function of binary responses and even where the underlying higher level distribution is Normal, we will need a reasonably large number of these responses to approximate it adequately, especially with very small or very large probabilities.

Table 4.3 Contraceptive use in Bangladesh. Standard errors in parentheses

	A (MQL1)	B (PQL2)
Fixed parameter		
Intercept	−1.59	−1.72
Age*	−0.025 (0.008)	−0.026 (0.008)
Urban	0.754 (0.162)	0.816 (0.170)
1 child	1.06 (0.16)	1.14 (0.16)
2 children	1.27 (0.17)	1.36 (0.18)
3+ children	1.26 (0.18)	1.36 (0.18)
Random parameter		
σ^2_{u0}(intercept)	0.334 (0.103)	0.396 (0.118)
σ_{u01}	−0.353 (0.142)	−0.414 (0.160)
σ^2_{u1} (urban)	0.596 (0.258)	0.685 (0.284)

* Age is centred at the mean age of 29.56. The base categories for other variables are 'rural' and 0 children. RIGLS estimation.

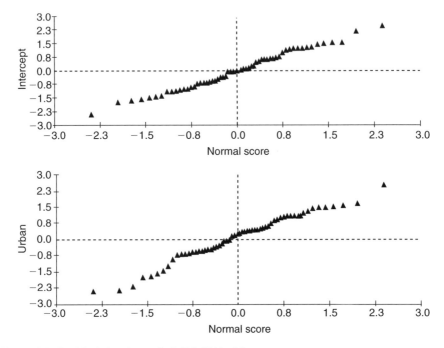

Figure 4.1   Residual plots for analysis B in Table 4.3.

Table 4.4   Estimates for model of Table 4.3 using alternative estimation procedures. Standard errors in parentheses (see text)

	Maximum likelihood (simulated)	MCMC	Iterated bootstrap
Fixed parameter			
Intercept	−1.72	−1.72	−1.71
Age*	−0.026 (0.008)	−0.027 (0.008)	−0.027 (0.009)
Urban	0.820 (0.170)	0.805 (0.189)	0.813 (0.182)
1 child	1.138 (0.160)	1.157 (0.157)	1.122 (0.153)
2 children	1.360 (0.177)	1.376 (0.174)	1.356 (0.169)
3+ children	1.358 (0.183)	1.384 (0.180)	1.354 (0.178)
Random parameter			
σ_{u0}^2(intercept)	0.390 (0.120)	0.418 (0.137)	0.403 (0.113)
σ_{u01}	−0.402 (0.182)	−0.432 (0.176)	−0.426 (0.164)
σ_{u1}^2 (urban)	0.654 (0.362)	0.738 (0.303)	0.714 (0.300)
Deviance criterion for addition of urban random effect. (DIC for MCMC)	15.8	22.6	

* See notes to Table 4.3.

Table 4.4 shows the parameter estimates for the same model using MCMC, maximum likelihood and the iterated bootstrap (see Appendix 4.4). The maximum likelihood estimates are obtained using simulated maximum likelihood with 400 simulated replicates. The MCMC estimation uses 5000 iterations and assumes flat diffuse priors for the fixed effects and minimally informative Wishart priors (using MQL1 random parameter estimates) for the random effects at level 2 and gamma priors at level 1 (see Appendix 2.4). The iterated bootstrap uses 500 replicates per bootstrap set and 10 sets.

The maximum likelihood deviance criterion for the urban random effects is tested as a chi-squared on two degrees of freedom and is rather larger than the Wald test for PQL2 estimates. For MCMC the DIC (see Chapter 2) is also substantial, indicating an improvement from adding urban random effects. The parameter estimates for all three methods are similar, and in particular, the variance estimates are somewhat larger than those for PQL2 and substantially larger than for MQL1.

4.4 Models for multiple response categories

In this section we extend the model for a single proportion as outcome to the case of a set of proportions, for example the proportions voting for three different political parties. The response is now multivariate and we can define a generalization of the ordinary logit model to define a multivariate logit as follows for a simple 2-level variance components model

$$\log\left(\frac{\pi_{ij}^{(s)}}{\pi_{ij}^{(t)}}\right) = \beta_0^{(s)} + \beta_1^{(s)} x_{ij} + u_j^{(s)}, \quad s = 1, \ldots, t-1 \tag{4.7}$$

where there are t response categories. Choosing one category (t) as the base category avoids redundancy and a singular covariance matrix and hence the need to introduce generalized inverses into the estimation. There are cases where this procedure is inappropriate and we discuss these below. Thus (4.7) specifies the model for each of the remaining $t-1$ categories with $\sum_{h=1}^{t} \pi_{ij}^{(h)} = 1$. When $t = 2$ this reduces to the ordinary logit model. We can also define a multivariate complementary log-log link as shown in Table 4.1 which again reduces to the ordinary log-log model when $t = 2$.

We treat the $t-1$ response categories as a multivariate response vector (see Chapter 6 for details about how multivariate response models are specified). In this case we use dummy variables with no variation at level 1 and the true level 1 covariance matrix specified at level 2. Consider, for example, a single-level model where individuals have three response categories, $t = 3$. We specify a bivariate model where level 2 describes the between-individual variation. If we make the standard assumption that the observed response proportions follow a multinomial distribution then the level 2 covariance matrix has the form

$$n_{ij}^{-1}\begin{pmatrix} \pi_{ij}^{(1)}(1 - \pi_{ij}^{(1)}) & & \\ -\pi_{ij}^{(1)}\pi_{ij}^{(2)} & \cdot & \\ \cdot & \cdot & \\ \cdot & \cdot & \cdot \\ -\pi_{ij}^{(1)}\pi_{ij}^{(t-1)} & \cdot \cdot \cdot & \pi_{ij}^{(t-1)}(1 - \pi_{ij}^{(t-1)}) \end{pmatrix} \tag{4.8}$$

where n_{ij} is the total number of responses over all categories.

We can create the covariance structure (4.8) as follows. Define the explanatory variables

$$z_{1ij} = \sqrt{\pi_{ij}/n_{ij}}, \quad z_{2ij} = \pi_{ij}/\sqrt{2n_{ij}}$$
$$z_{3ij} = -\pi_{ij}/\sqrt{2n_{ij}}, \quad \pi_{ij} = \{\pi_{ij}^{(s)}\} \tag{4.9}$$

and specify Z_1 to have a random coefficient at level 1 with variance constrained to 1.0 and Z_2, Z_3 to have random coefficients at level 2 constraining their variances to zero and their covariance to 1.0. This produces the structure (4.8) and extra multinomial variation can be achieved by allowing the variance and covariance to be different from 1.0 but constraining them to be equal. Level 3 then defines variation between higher level units, for example schools.

The response vector itself is not restricted to a single classification. Thus, suppose we are studying individual grades in examinations. If we had two responses each with three categories this produces nine response categories of which just one contains the value 1 for each individual. A 'main effects' model extension to (4.7) would express the probability of any particular combination of first and second preferences as an additive function of a term for the first and for the second exam grade as follows

$$\log\left(\frac{\pi_{ij}^{(s=s_1,s_2)}}{\pi_{ij}^{(t)}}\right) = \beta_0^{(s_1)} + \beta_0^{(s_2)} + \beta_1^{(s_1)}x_{1ij} + \beta_1^{(s_2)}x_{2ij} + u_j^{(s_1)} + u_j^{(s_2)}, \quad s = 1,\ldots,t-1$$

For the random parameters it might be reasonable to attempt to fit a model where the covariances between the $u_j^{(s_1)}$, $u_j^{(s_2)}$ were zero in order to reduce the number of random parameters in the model.

To see how we can interpret the parameters of these models we write, from (4.7)

$$\log(\pi_{ij}^{(r)}/\pi_{ij}^{(s)}) = (\beta_0^{(r)} - \beta_0^{(s)}) + (\beta_1^{(r)} - \beta_1^{(s)})x_{ij} + (u_j^{(r)} - u_j^{(s)}) \tag{4.10}$$

so that a unit change in x_{ij} multiplies the ratio of the r-th and s-th response probabilities by $\exp(\beta_r^{(r)} - \beta_1^{(s)})$. Likewise a difference of d in the residuals or in the intercept terms multiplies this ratio by e^d.

This formulation of the multicategory response model is adequate for models such as (4.7) where coefficients are fitted for each response category (except the base). There are other models, however, where we may wish to fit a function defined *across the categories*. This will often be the case when there are a large number of ordered categories where we wish to study linear, quadratic, etc. trends across the categories, although as we point out later there will often be more satisfactory procedures for such cases based upon consideration of the *cumulative* probabilities $\pi_{ij}^{(1)}$, $\pi_{ij}^{(1)} + \pi_{ij}^{(2)},\ldots$.

Where we do wish to treat the categories symmetrically and define a function across the response categories we replace the intercept term $\beta_0^{(s)}$ in (4.7) by such a function. If we assume a linear function then (4.7) can be written as

$$\log\left(\frac{\pi_{ij}^{(s)}}{\pi_{ij}^{(t)}}\right) = \gamma_0 + \gamma_1 w^{(s)} + (\beta_0 + \beta_1 w^{(s)})x_{ij} + u_j^{(s)}, \quad s = 1,\ldots,t-1 \tag{4.11}$$

where $w^{(s)}$ is the score assigned to category s. We might also wish to structure the level 2 variation, for example writing $u_j^{(s)} = u_{0j} + u_{1j}w^{(s)}$. Such a model will be especially useful where the number of categories becomes large.

In (4.11) the choice of base category is no longer irrelevant since the score assigned to this category does not appear in the model. We can avoid this difficulty by defining the multivariate logistic over all the response categories ($s = 1, \ldots, t$) and in (4.11) the level 2 resulting covariance matrix will not be singular so long as the set of response category probabilities is predicted using fewer responses than there are categories. An alternative formulation, using the Poisson with a log-link function as described below, will often be more convenient.

As with the case of a single proportion as outcome we have a choice of quasilikelihood, maximum likelihood or MCMC estimation procedures. With multiple responses, however, the biases associated with MQL can be very large for (0,1) responses and are also affected by choice of base category.

4.5 Models for counts

Instead of using a set of proportions as the response we can consider the underlying event counts as the set of responses. Thus, for example in the single exam grade case, suppose we classify individuals by three initial achievement categories and by gender. In each of the six cells within each level 2 unit, which is now, say, a school, we have counts of the numbers in each of the three grades, which yields 18 counts. The expected number of individuals in each grade can be written

$$m_{sij} = M_j\pi_{ij}^{(s)}$$

where s indexes the grades, i indexes the six cells within each level 2 unit and M_j is the number of individuals in the j-th school. Our inferences are therefore conditional on these totals. We write corresponding to (4.7)

$$\log(m_{sij}) = \log(M_j) + \beta_0^{(s)} + \beta_1^{(s)}x_{ij} + u_j^{(s)}, \quad s = 1, \ldots, t \quad (4.12)$$

The term $\log(M_j)$ is a fixed part offset and when using such offsets it may be better to centre them about their mean in order to avoid numerical instabilities. Corresponding to the multinomial assumption we now make the assumption of a Poisson distribution for the observed counts n_{sij}, which are assumed conditionally independent with

$$E(n_{sij}) = m_{sij}, \quad \text{var}(n_{sij}|m_{sij}) = m_{sij}$$

We now have a 2-level model where at level 2 we have the school and the level 1 units are the set of counts for the classification of grade by initial achievement category and gender. A basic additive model will have explanatory variables consisting of an intercept, two dummy variables for grade, two dummy variables for initial achievement and one for gender. We would normally also wish to include interactions between grade and initial achievement and grade and gender.

The level 1 variation is specified using the predicted number for each level 1 unit and the estimation follows the same pattern as for the binomial model, using the corresponding expressions given in Appendix 4.1. The level 1 random part will be defined by a dummy variable equal to the square root of the predicted count and with variance constrained to one where a Poisson distribution is assumed.

There are some applications where the response is a count and we do not require an offset, or where the offset is effectively constant. For example, if we were interested in the number of times individuals visited their general practitioner or physician in a year, we could collect data over a 1-year period for all individuals and study the variation in counts across practitioners (level 2) according to individual and practitioner characteristics.

There are variations on the Poisson distribution assumption which we may wish to use. For example, the negative binomial distribution can be obtained from a process whereby the response is generated by counting the number of incidents for each level 1 unit and where, conditional on the fitted explanatory variables and higher level terms, the mean count for each level 1 unit has a gamma distribution with index v. This leads us to consider level 1 variance functions of the general form $k_1 m + k_2 m^2$, where $k_1 = 1$ gives the negative binomial distribution with $k_2 = 1/v$. We could add further terms or consider even a nonlinear function.

As with binomial distribution models we can use maximum likelihood (Appendix 4.2) or MCMC estimation (Appendix 4.3).

4.6   Ordered responses

In the exam grade example of the previous section we ignored the fact that the scale was ordered. Such response scales are common and sometimes are analysed by assigning scores and then treating them as if they were continuous. While this may often be satisfactory, there are situations, for example where the distribution is very skew, where such a procedure is questionable. One possible alternative, mentioned in the preceding section, is to assign scores to the categories of the response variable and then carry out an analysis based upon the multinomial or Poisson model, using all the response categories in the analysis. Such a procedure, however, typically relies on choosing a suitable scoring system, just as does the continuous response model. One possibility in these cases is to assign scores by minimizing a measure of between-unit disagreement as in correspondence analysis or dual scaling (Greenacre, 1984; Goldstein, 1987c). In this section we shall look at procedures which avoid the arbitrariness of assumptions involved when assigning numerical scores.

To exploit the ordering we can base our models upon the *cumulative* response probabilities rather than the response probabilities for each category. We define these as

$$E(y_{ij}^{(s)}) = \gamma_{ij}^{(s)} = \sum_{h=1}^{s} \pi_{ij}^{(h)}, \quad s = 1, \ldots, t-1 \tag{4.13}$$

where $y_{ij}^{(s)}$ are the observed proportions out of a total n_{ij} and s now indexes the ordered cumulative categories. If we assume an underlying multinomial distribution for the category probabilities the cumulative proportions have a covariance matrix given by

$$\text{cov}(y_{ij}^{(s)}, y_{ij}^{(r)}) = \gamma_{ij}^{(s)}(1 - \gamma_{ij}^{(r)})/n_{ij}, \quad s \le r \tag{4.14}$$

We can therefore fit models to these cumulative proportions (or counts conditional on a fixed total) in the same way as with the multinomial response vector, substituting the covariance matrix (4.14) for (4.8). A discussion of these and related models is given in McCullagh and Nelder (1989). A common model choice is the *proportional odds* model which uses a logit link namely

$$\gamma_{ij}^{(s)} = \{1 + \exp(-[\alpha^{(s)} + (X\beta)_{ij}])\}^{-1} \tag{4.15}$$

which implies that increasing values of the linear component are associated with increasing probabilities with increasing s. We also require $\alpha^{(1)} \leq \alpha^{(2)} \ldots \leq \alpha^{(t-1)}$.

Another choice is the *proportional hazards* model which uses a log-log link to give

$$\gamma_{ij}^{(s)} = \{1 - \exp(-[\alpha^{(s)} + (X\beta)_{ij}])\} \tag{4.16}$$

An important special case of these models is where the categories are ordered in time so that $\alpha^{(s)}$ can be modelled as a function of time, and satisfying the above order relationship among these parameters. Some choices would be

$$\alpha^{(s)} = \delta \log(t_s), \quad \alpha^{(s)} = \delta t_s \tag{4.17}$$

Such a model might be used in developmental studies where individuals pass through a set of time-ordered stages. In studies of children, for example, it is possible to identify 'milestones' of development through which children pass, starting with none until all have been passed when developmental 'maturity' is reached. A repeated measures study would count the number passed at each time point so yielding a cumulative proportion in relation to time and other covariates. We would then be able to fit a 2-level model with variation between individuals involving any of the parameters in (4.15), (4.16) or (4.17). In the extreme case with just a single milestone, these models are equivalent to the event history models we consider in Chapter 10.

Another example of longitudinal discrete response data is where, at each measurement occasion, we have a vector of ordered categorical responses and each individual in the study responds to one category. The cumulative response vector for each individual at each occasion then contains a zero for each response category less than the category to which the individual responds and a one for that category and each higher category. We can model the time dependence within the set of explanatory variables X, and we would normally wish to include the possibility of interactions between the $\alpha^{(s)}$ and time. In such a model the basic covariance structure given by (4.14) represents the between-occasion covariation. Thus, although the data structure is represented by level 1 as the categories, level 2 as occasion and level 3 as individual, the higher level variation is only estimated at level 3. This can be compared to the simple binary response model where the binomial response variance is that between occasions, and the structure defines occasion as level 1 and individual as level 2 since there is a single response for each occasion. We also note that similar considerations apply to all the multicategory response models, with higher level variation estimated at level 3 and above.

4.7 Mixed discrete–continuous response models

An extension of the multivariate models considered in Chapter 6 is where some of the responses are continuous and some are discrete. For example, in a repeated measures study we may have a response which is the (discrete) maturity stage that an individual has reached as well as continuous measurements such as height and weight. In some circumstances we may wish to treat, say, the maturity stage as the response, conditional on height and weight and further covariates, including age. In other situations, for example if we are interested in prediction systems, then we would wish to consider all the measurements as responses, conditional on covariates. In another example, suppose we have measurements on smoking habit, including whether someone smoked and if so at what rate. We can consider this as a bivariate response model where each individual has a binary response for whether or not they smoke and if they do a further response for the number smoked per day.

We shall develop the model for the case of individual smoking habits with one binary and one continuous response and then look at the more general case of several binary responses. The extension to several responses of each type is straightforward as is the extension to multicategory responses and count data.

As in the standard multivariate multilevel model we have no variation at level 1 and at level 2, that of the individual, indexed by i, we have a binomial variance associated with the smoking/no smoking response and a between-individual variance for the number smoked. The variance for the binary response is the usual binomial variance and that for the continuous response is a parameter to be estimated. At the higher level, indexed by j, the variances and covariances will be defined in the standard fashion. For a 2-level model with individuals nested within, say, households we write the model in two parts; for the binary response probability

$$\text{logit}(\pi_{ij}) = (X_1\beta_1)_{ij} + u_{1j}$$

$$y_{ij} \sim binomial(1, \pi_{ij})$$

and for the continuous response

$$y_{ij} = (X_2\beta_2)_{ij} + u_{2j} + e_{ij}$$

with

$$e_{ij} \sim N(0, \sigma_e^2), \quad \begin{pmatrix} u_{1j} \\ u_{2j} \end{pmatrix} \sim N(0, \Omega_u), \quad \Omega_u = \begin{pmatrix} \sigma_{u1}^2 & \\ \sigma_{u12} & \sigma_{u2}^2 \end{pmatrix} \tag{4.18}$$

In the general case an individual can have any combination of responses, as in the maturity example, and the individual level covariance will have the form of a (adjusted) biserial covariance $(1 - \hat{\pi}_{jk})\hat{\pi}_{jk}(\hat{y}_{1jk} - \hat{y}_{2jk})$, where $\hat{\pi}_{jk}$ is the estimated probability of a positive response, and $\hat{y}_{1jk}$, $\hat{y}_{2jk}$ are respectively the predicted values of the continuous response for a positive and negative binary response. We can fit this using an extra covariance term in the model at the individual level, constrained to have the above value. If we assume that $\hat{y}_{1jk} - \hat{y}_{2jk}$ is constant then we can fit this term by defining a further explanatory variable equal to the existing variable defining the binomial variation at the individual level, and fitting just a covariance term between this further explanatory variable and the existing binomial explanatory variable. This gives the required estimate of $\hat{y}_{1jk} - \hat{y}_{2jk}$. In the smoking example, since non-smokers do not have any number smoked, this covariance term does not exist.

A particular case of interest is where we have two responses which are proportions. Suppose, for example, that in an educational survey we know the proportion of students in each school who pass an English exam and also the proportion who pass a Mathematics exam, but we have no information about how many pass or fail one or both. In other words, for the 2×2 table containing the numbers in each pass/fail category we only have the numbers in the margins. The level 2 covariance, in terms of the predicted proportions, has the form $\hat{\pi}_{(11)jk} - \hat{\pi}_{(1)jk}\hat{\pi}_{(2)jk}$. If we are prepared to assume that $\hat{\pi}_{(11)jk}$ is a function of $\hat{\pi}_{(1)jk}$ and $\hat{\pi}_{(2)jk}$, say proportional to their product, then with estimates of the marginal probabilities available from the model, the level 2 covariance estimate allows us to obtain an estimate of the joint probability of success on both Mathematics and English for a given set of explanatory variables. Note that the procedure depends upon the assumption of binomial variation. Of course, if we had all the original information then we would fit a model where there was a response for each cell of the table. This approach may also be used where separate surveys are conducted within the same level 2 units and each one produces a proportion as response. If there is overlap between the samples used, then there will exist level 2 covariances, and if information about the detailed nature of the overlap is available it will be possible in principle to obtain estimates of the joint probabilities.

In the next section we look at an alternative model that allows us to transform covariances between categorical variables to covariances between continuous ones.

4.8   A latent variable model for binary and ordered responses

In many cases a binary response can be thought of as being derived by choosing a threshold from an underlying continuous distribution so that the response is 1 if above the threshold and 0 if below. This might arise, for example, in assigning pass/fail gradings for an examination based on a mark cut-off. Likewise a series of ordered grades may be derived from an underlying mark scale. Another example is in the categorization of illness severity where there is no actual underlying continuous scale but interest may lie in constructing one from observed categories. A similar situation may arise with attitude measurement.

Suppose that we have a variance components 2-level model for the underlying continuous variable written as

$$y_{ij} = (X\beta)_{ij} + u_j + e_{ij} \qquad (4.19)$$

and suppose a positive value occurs when $y_{ij} > 0$. We then have

$$\Pr(y_{ij} > 0) = \Pr(e_{ij} > -[(X\beta)_{ij} + u_j]) \qquad (4.20)$$

Now if we assume $e_{ij} \sim N(0, 1)$, (equivalent to fixing the lowest level variance to be binomial) the probability in (4.20) is

$$\int\limits_{-[(X\beta)_{ij}+u_j]}^{\infty} \phi(t)dt = \int\limits_{-\infty}^{[(X\beta)_{ij}+u_j]} \phi(t)dt$$

where $\phi(t)$ is the probability density function (pdf) of $N(0, 1)$, which is in fact just the probit link function. In Appendix 4.3 we show how we can estimate the parameters of this model where every random variable has a Normal distribution. This is done using MCMC, sampling from the underlying level 1 $N(0,1)$ distribution. In the multivariate case this allows us to estimate a correlation matrix at level 1. We can carry out similar estimation for ordered data. In many cases this will be a more satisfactory approach than that described in section 4.6, especially where there are mixtures of continuous and binary responses.

We have an analogous model for the logit link function. We have the logistic distribution with the following density function

$$f(x) = \frac{\exp(x)}{[1 + \exp(x)]^2} \tag{4.21}$$

which has zero mean and variance 3.290 and where the required cumulative function is obtained as

$$\int_{-Y}^{\infty} f(x)dx = [1 + \exp(-Y)]^{-1} \tag{4.22}$$

which is just the logit link function. In this case, however, there is no natural extension to multivariate data so this is less useful.

We can also model an underlying continuous distribution for the log-log link, although again this has no natural extension to multivariate data. We have a Gumbel distribution with density function

$$f(x) = \exp(-x)(\exp -[\exp(-x)]) \tag{4.23}$$

which has mean -0.577 and variance 1.645. We have

$$\int_{-Y}^{\infty} f(x)dx = 1 - \exp(-\exp(Y)) \tag{4.24}$$

which is just the complementary log-log link function.

4.9   Partitioning variation in discrete response models

In Chapter 3 we pointed out that in random coefficient models the variance partition coefficient (VPC) was a function of the predictor variables with random coefficients. In discrete response models the computation and interpretation of the VPC is more complex. We discuss a model with a binary response, but our remarks will apply more generally to models for proportions, for different non-identity link functions and also where the response is a count, in fact to any non-linear model. For a (0,1) response, model (4.1) can be written

$$E(y_{ij}) = \pi_{ij} = f(\beta_0 x_0 + \beta_1 x_{1ij} + u_{0j})$$
$$y_{ij} \sim Bernoulli(\pi_{ij})$$
$$u_{0j} \sim N(0, \sigma_{u0}^2) \tag{4.25}$$

Unlike in the Normal case the level 1 variance depends on the expected value, $\text{var}(y_{ij}) = \pi_{ij}(1 - \pi_{ij})$ and the fixed predictor in the model depends on the value of x_1. Therefore as we are considering a function of the predictor variable x_1, a simple VPC is not available, even though there is only a single level 2 variance. Furthermore, the level 2 variance, σ_{u0}^2, is measured on the logistic scale so is not directly comparable to this level 1 variance.

If we still wish to produce a measure, however, the following procedures will provide at least approximate estimates.

Model linearization (Method A)

Using a first order Taylor expansion as described in Appendix 4.1 we can write (4.25) in the form

$$y_{ij} = (\beta_0 + \beta_1 x_{1ij}) f_{ij}' + u_j f_{ij}' + e_{ij} \sqrt{\pi_{ij}(1 - \pi_{ij})}$$

$$\text{var}(e_{0ij}) = 1$$

where we evaluate π_{ij} at the mean of the distribution of the level 2 random effect, that is, for the logistic model

$$\pi_{ij} = \exp(\beta_0 + \beta_1 x_{1ij})[1 + \exp(\beta_0 + \beta_1 x_{1ij})]^{-1}$$

$$f_{ij}' = \pi_{ij}[1 + \exp(\beta_0 + \beta_1 x_{1ij})]^{-1} \tag{4.26}$$

so that, for a given value of x_1 we have

$$\text{var}(y_{ij} | x_{1ij}) = \sigma_{u0}^2 \pi_{ij}^2 [1 + \exp(\beta_0 + \beta_1 x_{1ij})]^{-2} + \pi_{ij}(1 - \pi_{ij})$$

and

$$\tau = \sigma_{u0}^2 \pi_{ij}^2 [1 + \exp(\beta_0 + \beta_1 x_{1ij})]^{-2} \{\sigma_{u0}^2 \pi_{ij}^2 [1 + \exp(\beta_0 + \beta_1 x_{1ij})]^{-2} + \pi_{ij}(1 - \pi_{ij})\}^{-1}$$

where sample estimates are substituted.

Simulation (Method B)

This method is general and can be applied to any non-linear model without the need to evaluate an approximating formula. It is computationally reasonably fast and can be made to yield as accurate a result as desired by increasing the number of simulations. It consists of the following steps:

1. From the fitted model (say (4.25)) simulate a large number m (say 5000) values for the level 2 residual from the distribution $N(0, \sigma_{u0}^2)$, using the sample estimate of the variance.
2. For a particular chosen value(s) of x_1 compute the m corresponding values of $\pi_{ij}(\pi_{ij}^*)$ using (4.25). For each of these values compute the level 1 variance $v_{1ij} = \pi_{ij}^*(1 - \pi_{ij}^*)$.
3. The coefficient is now estimated as

$$\tau = v_2 (v_2 + v_1)^{-1}$$

$$v_2 = \text{var}(\pi_{ij}^*), \quad v_1 = E(v_{1ij})$$

A binary linear model (Method C)

As a very approximate indication for the VPC we can consider treating the (0,1) response as if it were a Normally distributed variable and estimate the VPC as in that case. This will generally be acceptable when the probabilities involved are not extreme, but if any of the underlying probabilities are close to 0 or 1, this model would not be expected to fit well, and may predict probabilities outside the (0,1) range.

A latent variable approach (Method D)

As in section 4.7 we may consider the observed (0,1) responses as arising from an underlying continuous variable so that a 1 is observed when a certain threshold is exceeded, otherwise a 0 is observed. For the logit model we have the underlying logistic distribution given by (4.21) and (4.22).

Since the variance for the standard logistic distribution is $\pi^2/3 = 3.29$ we take this to be the level 1 variance and both the level 1 and level 2 variances are on a continuous scale. We now simply calculate the ratio of the level 2 variance to the sum of the level 1 and level 2 variances to obtain the VPC. A similar computation can be used for the probit and log-log link functions.

This approach may be reasonable where the (0,1) response is, say, derived from a truncation of an underlying continuum such as a pass/fail response based upon a continuous mark scale, but would seem to have less justification when the response is truly discrete, such as mortality or voting. See also Snijders and Bosker (1999, Chapter 14) for a further discussion.

If we have fitted multiplicative extra-binomial variation (see section 4.1) then in the above approaches the level 1 variance is multiplied by the estimated scaling parameter. For additive extra-binomial variation this variance is added to the level 1 variance.

4.9.1   An example

We illustrate the procedures using some data on voting patterns (Heath et al., 1996). The model response is whether or not, in a sample of 800 respondents, they expressed a preference for voting Conservative ($y_{ij} = 1$) in the 1983 British general election. Several covariates were originally fitted but for simplicity we fit only an intercept model, namely

$$E(y_{ij}) = \pi_{ij} = \exp(\beta_0 + u_{0j})[1 + \exp(\beta_0 + u_{0j})]^{-1}$$

$$y_{ij} \sim Bernoulli(\pi_{ij})$$

$$u_{0j} \sim N(0, \sigma_{u0}^2)$$

The fitted model parameters are as follows (using PQL2 with IGLS):

$$\hat{\beta}_0 = -0.256, \quad \hat{\sigma}_{u0}^2 = 0.142$$

The VPC estimates are given in Table 4.5.

There is, as expected, good agreement for methods A, B and C with that for D somewhat larger. Now, however, we choose an extreme value. In chapter 8 of Rasbash et al. (2000) these data are fitted with four predictor variables measuring political attitudes. The scales of these predictor variables are constructed such that lower values

Table 4.5   Level 2 and level 1 estimated variances with variance partition coefficient (VPC)

	Method			
	A	*B*	*C*	*D*
Level 2 variance	0.0086	0.0083	0.0088	0.142
Level 1 variance	0.216	0.238	0.237	3.290
VPC	0.034	0.034	0.036	0.043

correspond to left-wing attitudes. If we take the set of values corresponding to the 5th percentile point of each of the four predictors, then the predicted value on the logit scale is approximately -2.5, corresponding to a low probability of voting Conservative of 0.076. At this value of the predictor the respective VPCs for methods A and B are 0.0096 and 0.0111, which are again in reasonable agreement. For method C, fitting the four predictor variables, we obtain a value of 0.0257 which is very different. For method D the residual level 2 variance does not change very much from the previous model, and we obtain a VPC of 0.047, which is also different to the others. We note that the VPC values for methods C and D do not depend on the value of the linear predictor.

If one wishes to make inferences on an underlying continuous scale then method D is appropriate. The choice of whether to report on the probability scale or an underlying continuous scale will depend on the application; in the present example it seems natural to report directly on the probability scale rather than on an assumed underlying continuous scale of 'propensity' to vote Conservative. We can obtain an interval estimate for any of our estimates via the bootstrap or using the results from an MCMC estimation run. Goldstein et al. (2002a) give more details of these procedures, including MLwiN macros for computing them.

Appendix 4.1

Generalized linear model estimation

4.1.1 Approximate quasilikelihood estimates

We consider a single nonlinear term of the form

$$y_{ij} = f(X_{2ij}\beta_2 + Z_{2ij}u_{2j}) + Z_{1ij}e_{2ij} \qquad (4.1.1)$$

At the $(t + 1)$-th iteration we expand the nonlinear function f in (4.1.1) for both fixed and random parts as follows using a Taylor series

$$f_{ij}(H_t) + X_{2ij}(\beta_{2,t+1} - \beta_{2,t})f'_{ij}(H_t) + (Z_{2ij}\,u_{2j})f'_{ij}(H_t)$$
$$+ (Z_{2ij}\,u_{2j})^2 f''_{ij}(H_t)/2 \qquad (4.1.2)$$

in terms of parameter values estimated at the t-th iteration. The first line of (4.1.2) updates the fixed part of the model and in the special case of a single-level model provides the usual updating function in maximum likelihood estimation. The quantity $f_{ij}(H_t) - X_{2ij}\beta_{2,t}f'_{ij}(H_t)$ is treated as an offset to be subtracted from the response variable. The first term in the second line defines a linear random component based on the explanatory variables transformed by multiplying by the first differential. We need to specify H_t and then consider the distribution of the second term in the second line of (4.1.2). The level 1 term in (4.1.1) is fitted in the usual way; the choice of Z_{1ij} depending on the nature of the response as discussed in Section 4.1.

If we choose $H_t = X_{2ij}\beta_{2,t}$, this is equivalent to carrying out the Taylor expansion around the fixed part predicted value. If we choose $H_t = X_{2ij}\beta_{2,t} + Z_{2ij}\hat{u}_{2j}$, this expands around the current predicted value for the ij-th unit and we replace the second line of (4.1.2) by

$$Z_{2ij}(u_{2j} - \hat{u}_{2j})f'_{ij}(H_t) + Z_{2ij}(u_{2j} - \hat{u}_{2j})^2 f''_{ij}(H_t)/2$$

We thus have the further offset from the linear term to be added to the response

$$(Z_{2ij}\hat{u}_{2j})f'_{ij}(H_t)$$

A discussion of these approaches in the context of multilevel generalized linear models is given by Breslow and Clayton (1993). Wolfinger (1993) synthesizes some of the literature based upon this 'predictive' approach. All these methods use only the first order terms in (4.1.2).

From the second line of (4.1.2), where the Taylor expansion is about zero, we have

$$E(Z_{2ij}u_{2j}) = 0, \quad E(Z_{2ij}u_{2j})^2 = \sigma_{zu}^2, \quad \sigma_{zu}^2 = Z_{2ij}\Omega_u Z_{2ij}^T \tag{4.1.3}$$

To incorporate the second order terms we treat $\sigma_{zu}^2 f''(H_t)/2$ as an additional offset in the fixed part and in the random part of the model we need to consider the variation of the second term in the second line of (4.1.2). If we assume Normality then all third moments, formed from the product of the two terms in the second line of (4.1.2), are zero and we have

$$\text{var}(Z_{2ij}u_{2j})^2 = 2(\sigma_{zu}^4) \tag{4.1.4}$$

so that we need to define the additional random variable $Z_u^* = \sigma_{zu}^2 f''(H_t)/\sqrt{2}$ with variance constrained to be equal to 1.0. Equivalently we can form $Z_u^* Z_u^{*T}$ as an offset for the response vector $vec(\tilde{Y}\tilde{Y}^T)$ in the estimation of the random parameters.

Having modified the response variable by removing the necessary offsets we are left in the fixed part with a modified response, say Y', with a modified explanatory variable matrix, say X'. We do likewise for the random part of the model and then carry out a standard iterative procedure, updating the differential functions at each iteration.

Where the Taylor expansion is taken about the current values of the residuals we require $E[Z_{2ij}(u_{2j} - \hat{u}_{2j})]^2$ which leads to the 'conditional' or 'comparative' variances of the residuals and these are substituted as described in Appendix 2.2, so that we substitute these in the above expressions for the fixed and random offsets.

To estimate residuals we note that, having adjusted the response using the offsets, we have on the right-hand side of the model, for the Taylor expansion about zero, the fixed part together with the random terms

$$(Z_{2ij}u_{2j})f_{ij}'(H_t) + [(Z_{2ij}u_{2j})^2 - (\sigma_{zu}^2)]f_{ij}''(H_t)/2$$

Each residual and its square appear in this expression, and since third order moments are zero, we can apply the usual linear estimation for the residuals as described in Appendix 2.2. The weight matrix V is based upon both the linear and quadratic terms of the above expression. We carry out an analogous procedure for the case where the Taylor expansion is based upon the current residual estimates.

The above can be extended in a straightforward way to more than two levels and of course to multivariate models. For the first order approximation the procedure outlined here is closely related to that given by Lindstrom and Bates (1990) for 2-level repeated measures data who consider a first order expansion about the unit-specific predicted values. Gumpertz and Pantula (1992) consider a variance components model where the fixed part predictor is nonlinear.

For generalized linear models Waclawiw and Liang (1993) consider a generalized estimating equations approach (see Chapter 2), using a unit-specific predictor. A full likelihood based method for a repeated measures model with binary responses is described by Garret et al. (1993), and issues arising from applications to repeated measures data are discussed in Chapter 9.

For small samples, as discussed in Appendix 2.1, we should use the (RIGLS, REML) procedure. As discussed in Chapter 4, for certain data configurations this procedure can produce biased estimates of the fixed and random parameters. In Appendix 4.2 we discuss maximum likelihood estimation and bootstrapping.

4.1.2   Differentials for some discrete response models

The Logit–Binomial model

$$f = [1 + \exp(-X\beta)]^{-1}$$
$$f' = f[1 + \exp(X\beta)]^{-1}$$
$$f'' = f'[1 - \exp(X\beta)][1 + \exp(X\beta)]^{-1}$$

The Logit–Multinomial (Multivariate Logit) model

$$f^{(s)} = \exp(X\beta^{(s)}) \left[1 + \sum_{h=1}^{t-1} \exp(X\beta^{(h)}) \right]^{-1}, \quad s = 1, \ldots, t-1$$

$$f'^{(s)} = f^{(s)} \left[1 + \sum_{h \neq s} \exp(X\beta^{(h)}) \right] \left[1 + \sum_{h=1}^{t-1} \exp(X\beta^{(h)}) \right]^{-1}$$

$$f''^{(s)} = f'^{(s)}(1 - 2f^{(s)})$$

The Log–Poisson model

$$f = \exp(X\beta)$$
$$f' = \exp(X\beta)$$
$$f'' = \exp(X\beta)$$

The Log-log–Binomial model

$$f = 1 - \exp[-\exp(X\beta)]$$
$$f' = (1 - f)\exp(X\beta)$$
$$f'' = f'[1 - \exp(X\beta)]$$

Appendix 4.2

Maximum likelihood estimation for generalized linear models

4.2.1 Simulated maximum likelihood estimation

We have seen that for Normally distributed data maximum likelihood (ML) or restricted maximum likelihood (REML) estimates can be obtained readily using the IGLS and RIGLS algorithms. For discrete response or generalized linear models the estimation becomes more complicated. We discuss below how numerical quadrature can be used but that this becomes infeasible for models with large numbers of random coefficients and/or levels. Several simulation-based alternatives have been suggested (see for example McCulloch (1997) who gives an example for a logit model). One of these methods uses a stochastic version of the EM algorithm where, instead of obtaining the relevant expectations in the E step, the expectations are estimated by generating draws from the estimated distribution of the random effects using an MH algorithm with a suitable proposal distribution and averaging over the draws. Another method is to use an iterative (scoring) algorithm for a generalized linear model where for the weights in the estimation of updated parameters we use expectations calculated again as averages over MH sampled values.

A useful general procedure, which is relatively straightforward to implement, is simulated maximum likelihood where the likelihood is evaluated directly by simulation. We can write the full likelihood and corresponding log-likelihood as

$$L(\beta, \Omega, U) = \prod f(Y|U; \beta) f(U; \Omega)$$
$$\log[L(\beta, \Omega, U)] = \sum \{\log[f(Y|U; \beta)] + \log[f(U; \Omega)]\}$$

(4.2.1)

Lee and Nelder (1996) refer to this joint likelihood as the *h-likelihood* and this leads to the estimation methods of Appendix 4.1 based upon approximating the nonlinear function $f(Y|U; \beta)$. Lee and Nelder (2001) provide a generalization that covers a wide class of models with different distributional assumptions and variance structures.

An exact likelihood approach is to estimate the fixed and random parameters of our model by treating the actual random effects U in (4.2.1) as nuisance parameters, and to integrate them out and work with the *marginal likelihood* given by

$$L(\beta, \Omega) = \int f(Y|U; \beta) f(U; \Omega) dU$$

(4.2.2)

where the first term on the right-hand side is the distribution function for the responses conditional on the random effects, or residuals, U. The second term is the distribution function for the random effects. The first term, given U, depends only on the unknown parameters β and the second only on the unknown parameters Ω. Thus, for example, for a 2-level logistic binomial response model where the random effects are assumed to be multivariate normal we have, since the random effects are independent across units,

$$L(\beta, \Omega) = \prod_j \int \prod_i \{(\pi_{ij})^{s_{ij}}(1 - \pi_{ij})^{n_{ij}-s_{ij}}\}\Phi(u_j; \Omega)du_j$$

$$\pi_{ij} = \{1 + \exp(-X_{ij}\beta_j)\}^{-1}, \quad \beta_j = \beta + u_j$$

(4.2.3)

where Φ is the multivariate normal density function for the u_j and n_{ij}, s_{ij} are the numbers of trials and successes.

Now (4.2.2) is simply the expected value of $f(Y|U; \beta)$ over the distribution of the random effects U. For any given choice of β, Ω we could therefore, in principle, approximate the likelihood by repeatedly sampling sets of random effects from $f(U; \Omega)$ computing $f(Y|U; \beta)$ and forming

$$\sum_{h=1}^{N} f(Y|U_h; \beta)/N$$

(4.2.4)

where N is a suitably large number to achieve an acceptable accuracy. The problem is that we do not know the maximum likelihood values of the random parameters to use in the sampling. Suppose, however, that we have a good approximation to Ω, say $\tilde{\Omega}$ and we write (4.2.2) as

$$L(\beta, \Omega) = \int \frac{f(Y|U; \beta)f(U; \Omega)f(U; \tilde{\Omega})}{f(U; \tilde{\Omega})} dU$$

(4.2.5)

We can approximate (4.2.5) for the j-th unit by

$$\sum_{h=1}^{N} \frac{f(Y|U_j; \beta)f_h(U_j; \Omega)}{f_h(U_j; \tilde{\Omega})} \Big/ N$$

(4.2.6)

where sampling is now with respect to the (importance) distribution of the random effects given the value $\tilde{\Omega}$. The term

$$\frac{f(U; \tilde{\Omega})}{f(U; \Omega)}$$

adjusts for the different probability density of the importance distribution.

This is known as *importance sampling* and (4.2.6) will provide the required likelihood to any degree of accuracy for a choice β, Ω, for any choice of $\tilde{\Omega}$, although a poor choice will not be efficient. We can improve efficiency by choosing $f(U; \tilde{\Omega})$ so that $U_j \sim N(\hat{U}_j, \hat{\Omega}_{uj})$, where $\hat{U}_j$ is the set of estimated residuals for unit j with covariance matrix Ω_{uj}. These can be estimated from a preliminary sampling run based upon the initial MQL or PQL starting values (see below).

In practice it is more convenient to work with the log-likelihood so that, for example, for the binomial logit–normal model the log-likelihood is

$$
L = \sum_{j} \left(\ln \left\{ \sum_{h=1}^{N} \left[\prod_{i} \left\{ (\pi_{ij}^{(h)})^{s_{ij}} (1 - \pi_{ij}^{(h)})^{n_{ij} - s_{ij}} \right\} \frac{\Phi(u_j^{(h)}; \Omega)}{\Phi(u_j^{(h)}; \tilde{\Omega})} \right] / N \right\} \right)
$$

$$
= \sum_{j} \left(\ln \left\{ \sum_{h=1}^{N} \exp \left[\sum_{i} [s_{ij} \ln(\pi_{ij}^{(h)}) + (n_{ij} - s_{ij}) \ln(1 - \pi_{ij}^{(h)})] \right. \right. \right. \tag{4.2.7}
$$

$$
\left. \left. \left. + \ln \left(\frac{\Phi(u_j^{(h)}; \Omega)}{\Phi(u_j^{(h)}; \tilde{\Omega})} \right) \right] / N \right\} \right)
$$

$$
= \sum_{j} \left(\ln \left\{ \sum_{h=1}^{N} \exp \left[\sum_{i} [s_{ij} \ln(\pi_{ij}^{(h)}) + (n_{ij} - s_{ij}) \ln(1 - \pi_{ij}^{(h)})] \right. \right. \right.
$$

$$
\left. \left. \left. + \tfrac{1}{2} u_j^{(h)^T} (\tilde{\Omega}^{-1} - \Omega^{-1}) u_j^{(h)} + \tfrac{1}{2} \ln \left(\frac{|\tilde{\Omega}|}{|\Omega|} \right) \right] \right\} - \ln(N) \right)
$$

where h indexes the sets of random draws and u is a $p \times 1$ vector for each level 2 unit, where p is the number of random coefficients. Note that if $\Omega = 0$, the terms in Ω are omitted. To find the maximum likelihood solution near to $\tilde{\Omega}$ we need to carry out a numerical search over the parameter set β, Ω. Standard routines for maximizing general functions can be used and are available in various mathematical and statistical software packages. Some care is required to ensure that the calculations are carried out and stored with sufficient accuracy and double precision arithmetic will usually be necessary. When generating the U the random number seed should be reset to the same value at the start of each set of random draws. This avoids instability due to the stochastic nature of the likelihood estimates. Also, this stochastic element gives a small bias in the estimates of the parameters which decreases as N increases, and is an area for further research. For $\tilde{\Omega}$ we can use second order PQL (PQL2) estimates, or where these may be biased or unobtainable due to convergence problems we can use PQL1 or MQL estimates, possibly together with an iterated bootstrap, to obtain Ω. An approximation to the covariance matrix of the estimated parameters can be obtained from the inverse of the (Hessian) matrix of second derivatives of (minus) the log-likelihood with respect to the parameters at the solution point. This will be available as output from the optimization routine.

Note that the extension to models with more levels is relatively straightforward. Thus, for a 3-level model, since there is independence across levels, we have for the j-th level 2 unit within the k-th level 3 unit

$$
\Phi(u_{kj}^{(h)}, u_k^{(h)}; \Omega) = \Phi(u_k^{(h)}; \Omega_3) \Phi(u_{jk}^{(h)}; \Omega_2)
$$

$$
\Omega = \begin{pmatrix} \Omega_3 & \\ & \Omega_2 \end{pmatrix}
$$

and for the k-th level 3 unit the third line of (4.2.6) becomes

$$\sum_j \left(\ln \left\{ \sum_{h=1}^{N} \exp \left[\sum_i \left[s_{ijk} \ln\left(\pi_{ijk}^{(h)}\right) + (n_{ijk} - s_{ijk}) \ln\left(1 - \pi_{ijk}^{(h)}\right) \right] + \tfrac{1}{2} u_{jk}^{(h)^T} \right. \right. \right.$$

$$\left. \left. \left. \times \left(\tilde{\Omega}_2^{-1} - \Omega_2^{-1}\right) u_{jk}^{(h)} + \tfrac{1}{2} u_k^{(h)^T}\left(\tilde{\Omega}_3^{-1} - \Omega_3^{-1}\right) u_k^{(h)} + \tfrac{1}{2} \ln\left(\frac{|\tilde{\Omega}_2| \, |\tilde{\Omega}_3|}{|\Omega_2| \, |\Omega_3|} \right) \right] \right\} - \ln(N) \right)$$

and these are summed over level 3 units. At the maximum likelihood estimates we also have an estimate of the likelihood itself and this can be used for inference, for example for comparing two nested models. Likewise, a numerical search procedure can be extended to provide full or profile likelihood confidence intervals (Chapter 2).

This approach can be generalized to other distributions, link functions and structures. Thus, for example, for continuous responses we can fit a t-distribution at level 1 or, say, a beta distribution. Likewise various types of link function can be used, including the complementary log-log or probit for binomial data. Multinomial, ordered or unordered data can be fitted using a suitable parameterization. We can also fit more complex structures such as mixture distributions.

Residuals

The residuals at higher levels are defined as the expected values of the random effects given the data and parameter values. Thus we have

$$E(U) = \int U f(Y|U; \beta) f(U; \Omega) \, dU$$

and we obtain the expected value of u as

$$\exp \left[\sum_j \left(\ln \left\{ \sum_{h=1}^{N} \left[\prod_i \left\{ \left(\pi_{ij}^{(h)}\right)^{s_{ij}} \left(1 - \pi_{ij}^{(h)}\right)^{n_{ij} - s_{ij}} \right\} u_{j,k}^{(h)} \frac{\Phi(u_j^{(h)}; \Omega)}{\Phi(u_j^{(h)}; \tilde{\Omega})} \right] \right/ N \right\} \right) - L \right]$$

and the expected value of the product of two random effects $u_{k_1} u_{k_2}$ as

$$\exp \left[\sum_j \left(\ln \left\{ \sum_{h=1}^{N} \left[\prod_i \left\{ \left(\pi_{ij}^{(h)}\right)^{s_{ij}} \left(1 - \pi_{ij}^{(h)}\right)^{n_{ij} - s_{ij}} \right\} u_{j,k_1}^{(h)} u_{j,k_2}^{(h)} \frac{\Phi(u_j^{(h)}; \Omega)}{\Phi(u_j^{(h)}; \tilde{\Omega})} \right] \right/ N \right\} \right) - L \right]$$

from which we can obtain the estimated covariance matrix of the residuals. Note that we need to scale the expected values using the likelihood L.

Cross-classifications and multiple membership models

For a two-way cross-classification we have the log-likelihood

$$\ln \left\{ \sum_{h=1}^{N} \exp \left[\sum_j \sum_i \left[s_{ij} \ln\left(\pi_{ij}^{(h)}\right) + (n_{ij} - s_{ij}) \ln\left(1 - \pi_{ij}^{(h)}\right) \right] + \tfrac{1}{2} u_i^{(h)^T}\left(\tilde{\Omega}_A^{-1} - \Omega_A^{-1}\right) u_i^{(h)} \right. \right.$$

$$\left. \left. + \tfrac{1}{2} u_j^{(h)^T}\left(\tilde{\Omega}_B^{-1} - \Omega_B^{-1}\right) u_j^{(h)} + \tfrac{1}{2} \ln\left(\frac{|\tilde{\Omega}_A| \, |\tilde{\Omega}_B|}{|\Omega_A| \, |\Omega_B|} \right) \right] \right\} - \ln(N)$$

where i is associated with classification A and j with classification B. Note that we have to sum over all level 2 cells of the cross-classification for each value of h.

For a multiple membership model at level 2 we simulate from a multivariate Normal (or t) distribution as above, but now for a level 1 unit the level 2 covariance matrix is given by

$$\left(\sum_j w_{ij}^2 \right) \Omega_2$$

where the weights w are specific to the level 1 unit. Thus we form the multiple membership weighted sum $\sum_j w_{ij}u_j = u_{iw}$, and use these random effects applied to the above covariance matrix. Again we sum over all level 2 units for each h.

Computing issues

Several computing issues arise. The number of simulation sets N is a matter for further research, although values between 1000 and 2000 have proved adequate on the limited range of examples so far studied. The optimization procedure is also important and efficient nonlinear search algorithms are necessary. Inequality constraints are needed to ensure positive definite covariance matrices and upper and lower bounds for the fixed parameters may be important to ensure efficiency, as well as good starting values, for example from a PQL estimation. General search procedures are able to compute search directions numerically but supplying analytical first derivatives will improve speed and reliability. The first derivatives of L with respect to the parameters can be computed using the formulae in Appendix 4.1. Thus for the logit model we have the first derivative

$$\frac{\partial L}{\partial \theta} = \frac{1}{\sum_{h=1}^{N} \exp(g_j^{(h)})} \sum_j \frac{\partial}{\partial \theta} \sum_{h=1}^{N} \exp(g_j^{(h)})$$

$$= \frac{1}{\sum_{h=1}^{N} \exp(g_j^{(h)})} \sum_j \sum_h \left(\exp(g_j^{(h)}) \sum_i \left\{ \frac{\partial}{\partial \theta}[s_{ij} \ln(\pi_{ij}^{(h)}) + (n_{ij} - s_{ij}) \ln(1 - \pi_{ij}^{(h)})] \right\} \right.$$

$$\left. + \frac{\partial}{\partial \theta}\left[\frac{1}{2}u_j^{(h)T}(\tilde{\Omega}^{-1} - \Omega^{-1})u_j^{(h)} + \frac{1}{2}\ln\left(\frac{|\tilde{\Omega}|}{|\Omega|} \right) \right] \right)$$

where

$$g_j^{(h)} = \sum_i \left[s_{ij} \ln(\pi_{ij}^{(h)}) + (n_{ij} - s_{ij}) \ln(1 - \pi_{ij}^{(h)}) \right]$$

$$+ \frac{1}{2}u_j^{(h)T}(\tilde{\Omega}^{-1} - \Omega^{-1})u_j^{(h)} + \frac{1}{2}\ln\left(\frac{|\tilde{\Omega}|}{|\Omega|} \right)$$

For a fixed coefficient β_k this becomes

$$\frac{1}{\sum_{h=1}^{N} \exp(g_j^{(h)})} \sum_j \sum_h (\exp(g_j^{(h)}) \sum_i \frac{x_{ij,k}}{1 + \exp(X\beta)_{ij}^{(h)}} \{[s_{ij}(1 + \exp(X\beta)_{ij}^{(h)}) - n_{ij}\exp(X\beta)_{ij}^{(h)}]\}$$

and for variance and covariance terms we obtain corresponding expressions using

$$\frac{\partial}{\partial\theta}\left[\frac{1}{2}u_j^{(h)^T}(\tilde{\Omega}^{-1}-\Omega^{-1})u_j^{(h)}+\frac{1}{2}\ln\left(\frac{|\tilde{\Omega}|}{|\Omega|}\right)\right] = -\frac{\partial}{\partial\theta}\left[\frac{1}{2}u_j^{(h)^T}\Omega^{-1}u_j^{(h)}+\frac{1}{2}\ln(|\Omega|)\right]$$

$$=\frac{1}{2}u_j^{(h)^T}\Omega^{-1}\frac{\partial\Omega}{\partial\sigma_{kl}}\Omega^{-1}u_j^{(h)}-\frac{1}{2}tr\left(\Omega^{-1}\frac{\partial\Omega}{\partial\sigma_{kl}}\right)$$

where, for example for a bivariate distribution,

$$\frac{\partial\Omega}{\partial\sigma_1^2}=\begin{pmatrix}1 & 0\\0 & 0\end{pmatrix}, \quad \frac{\partial\Omega}{\partial\sigma_{12}}=\begin{pmatrix}0 & 1\\1 & 0\end{pmatrix}.$$

4.2.2 Maximum likelihood estimation via quadrature

We saw above that the likelihood for a 2-level logistic model can be written as the product of terms such as the following for level 2 unit j

$$\int_{-\infty}^{\infty}\prod_i\{(\pi_{ij})^{s_{ij}}(1-\pi_{ij})^{n_{ij}-s_{ij}}\}f(u_j;\Omega)du_j$$

$$\pi_{ij}=\{1+\exp(-X_{ij}\beta_j)\}^{-1}, \quad \beta_j=\beta+u_j \tag{4.2.8}$$

where $f(u_j;\Omega)$ is typically assumed to be the multivariate Normal density and can be written in the form $\int_{-\infty}^{\infty}P(u_j)f(u_j)\,du_j$.

Gauss–Hermite quadrature approximates an integral such as the above as

$$\int_{-\infty}^{\infty}P(v)e^{-v^2}\,dv\approx\sum_{q=1}^{Q}P(x_q)w_q \tag{4.2.9}$$

where the right-hand side is a Gauss–Hermite polynomial evaluated at a series of quadrature points indexed by q. Hedeker and Gibbons (1994) give a detailed discussion and also consider the multicategory (multinomial) response case. This function is then maximized using a suitable search procedure over the parameter space. Estimates produced from, for example, a PQL analysis will help to locate the search region.

If we consider the model with a single random intercept at level 2 we have

$$P(u_j)=\prod_i\frac{\exp(X_{ij}\beta+u_j)}{\{1+\exp(X_{ij}\beta+u_j)\}^2}, \quad f(u_j)=\sigma_u^2\phi \tag{4.2.10}$$

where ϕ is the standard Normal density. The standard quadrature method selects points centred around zero, but the u_j are not centred at zero and we may therefore need a very large number of quadrature points to cover the range. A solution is to use *adaptive quadrature*. One possibility (Liu and Pierce, 1994) is to centre the quadrature points on the modal (or mean) values of the function $P(u_j)$ and to scale them suitably, for example according to the estimated standard deviation of u_j (see Hartzel et al., 2001 for a discussion). These central values need to be estimated and a convenient choice is to use preliminary PQL estimates together with estimated standard errors.

Quadrature methods have been applied successfully to Poisson, binomial and multinomial and ordered category models and have been implemented in software packages (SAS, MIXOR, AML). Nevertheless, successful quadrature, even with the adaptive method, will often require a large number of quadrature points and even in simple cases convergence can be difficult to achieve (Lessaffre and Spicssens, 2001). This becomes especially important when there are several random coefficients, since the quadrature points will now be in several dimensions so that the number of points increases geometrically with the number of random coefficients. This places a practical limit on the complexity of models that can be handled in this way and applications have been mainly restricted to 2 or 3-level models with small numbers of random terms. Gauss–Hermite quadrature is effectively limited to the Normal distribution because of the exponential term in (4.2.9) and alternative quadrature methods are required for other higher level distributions.

Appendix 4.3

MCMC estimation for generalized linear models

4.3.1 MH sampling

In Appendix 2.4 the basic Gibbs sampling algorithm for models with Normal random effects was described. In the case of generalized linear models we cannot easily write down the conditional distribution for every step of the algorithm, for example for the fixed effects and the residuals. Thus, for the fixed effects step with a single level 2 variance for a binary response we have

$$p(\beta|y,\sigma_u^2,\sigma_e^2,u) \propto L(y;\beta,u,\sigma_e^2)p(\beta)$$

where

$$L(y;\beta,u,\sigma_e^2) = (1 + e^{-(X\beta+u)})^{-y}(1 + e^{-(X\beta+u)})^{y-1}$$

and a suitable diffuse prior is $p(\beta) \propto 1$.

We can therefore use MH sampling (section 2.4.2) and in general can mix Gibbs sampling with MH sampling steps. In other respects the estimation is as for Normal models with the corresponding likelihood functions for binomial, Poisson, multinomial, etc. data replacing the Normal likelihood (see Browne and Draper (2000) for more details).

4.3.2 Latent variable models for binary data

In the case of the probit link function for binomial data we can avoid the use of MH sampling and obtain an additional interpretation as follows. We consider a binary response where a positive response (=1) occurs when the value of an underlying continuous variable exceeds a threshold. For example, an examination may yield pass/fail grades which can be supposed to have such an underlying continuous scale. To illustrate write a variance components 2-level model for the underlying continuous variable as

$$y_{ij} = (X\beta)_{ij} + u_j + e_{ij} \tag{4.3.1}$$

and suppose a positive value occurs when $y_{ij} > 0$. We then have

$$\Pr(y_{ij} > 0) = \Pr(e_{ij} > -[(X\beta)_{ij} + u_j]) \tag{4.3.2}$$

Now if we assume $e_{ij} \sim N(0,1)$, (equivalent to fixing the lowest level variance to be binomial), the probability in (4.3.2) is

$$\int_{-[(X\beta)_{ij}+u_j])}^{\infty} \phi(t)dt, \quad \text{where } \phi(t) \text{ is the pdf of } N(0,1) \qquad (4.3.3)$$

In the standard Gibbs algorithm we insert an extra step which generates a random value y_{ij} from the truncated Normal distribution defined by (4.3.3) given current parameter values for each level 1 unit. When the response value is a 1 we select from $[-X^*, \infty]$, $X^* = (X\beta)_{ij} + u_j$ and where the response is 0 from $[-\infty, X^*]$. These are then used with (4.3.1) as in the standard Normal case. An advantage of this approach is that it is generally faster than using MH. Additionally, however, because it generates a set of 'imputed' continuous responses we can readily combine it with observed Normal responses, so allowing estimates of underlying correlations at level 1 between binary and continuous variables or between several binary variables. One application is to the fitting of binary response factor analysis models (see Chapter 7).

A similar approach can be used for the logit and log-log link functions, although these do not lead to multivariate Normal distributions where more than one response is involved and so are less useful.

For logit link models we need to take a random draw from the logistic distribution with the following density function

$$f(x) = \frac{\exp(x)}{[1 + \exp(x)]^2} \qquad (4.3.4)$$

which has zero mean and variance 3.290 and where the required cumulative function is obtained as

$$\int_{-Y}^{\infty} f(x)dx = [1 + \exp(-Y)]^{-1} \qquad (4.3.5)$$

which is the logit link function, so that we take a random draw as above, but this time from the truncated logistic distribution with truncation point corresponding to $X^* = (X\beta)_{ij} + u_j$ as before. To select a random draw from this logistic distribution we first take a random draw, u, from the uniform $(0, 1)$ distribution and then make the transformation $y = \log(u/(l - u))$. The corresponding 2-level model now has the level 1 variance constrained to 3.29.

For the log-log link we take a random draw from the Gumbel distribution with density function

$$f(x) = \exp(-x)(\exp - [\exp(-x)]) \qquad (4.3.6)$$

which has mean –0.577 and variance 1.645. We have

$$\int_{-Y}^{\infty} f(x)dx = 1 - \exp(-\exp(Y)) \qquad (4.3.7)$$

which is the complementary log-log link function and the transformation from the uniform is now $y = \log(-\log(1 - u))$ and we draw from the truncated distribution with truncation point given by $X^* = (X\beta)_{ij} + u_j$ and the level 1 variance constrained to 1.645.

4.3.3 Multicategory ordered responses

Consider the ordered category case with underlying cumulative proportions for categories $s = 1,..., p\text{-}1$ (omitting subscripts) and where each level 1 unit response vector consists of a set of zeros followed by a set of ones determined by the first (cumulative) category they respond to. We again assume an underlying continuum. Corresponding to (4.3.1) for a variance components model we write, omitting subscripts

$$Y^{(s)} = X\beta + u + \theta_s + e$$
$$e \sim N(0,1) \tag{4.3.8}$$

where the θ_s are parameters corresponding to the cut points on the assumed underlying continuum. For category s the probability of a positive (cumulative) response is given by

$$\Pr(e < -[(X\beta) + u + \theta_s]) \tag{4.3.9}$$

This reduces to (4.3.2) for two categories with $\theta_1 = 0$, where the cut point value is incorporated into the intercept term in $X\beta$. The cut point parameters θ_s can be modelled using $p\text{-}1$ dummy variables and we select from the corresponding truncated distributions. Note that $X\beta$ here does not include an intercept, although it may be convenient to specify one in which case there are $p\text{-}2$ remaining cut point parameters measuring deviations from the intercept term. As with binary data we can extend this to sets of binary or mixed binary and continuous responses. The logit and log-log link functions can be extended in analogous ways.

4.3.4 Proportions as responses

We can extend the (0,1) response model to the case where the response is a proportion. For the simple binomial probit model suppose there are r 'successes' out of n 'trials'. Assuming independent sampling we generate a set of r values y_{ij} where the response is 1, i.e. from $[-X^*, \infty]$, $X^* = (X\beta)_{ij} + u_j$, and $n\text{-}r$ from $[-\infty, X^*]$. We can think of this set of values as defining a lowest level cluster, essentially a replication level, below the original level 1 of the model and proceed to fit this modified model with an extra level where there is a single variance term constrained to be equal to 1. This leads to a simple modification to the Gibbs algorithm (see Appendix 2.4). Similar modifications can be made for the other models described above.

Appendix 4.4

Bootstrap estimation for generalized linear models

4.4.1 The iterated bootstrap

As pointed out in section 3.6.2, we can use the bootstrap to provide a bias correction for parameter estimates. This only works, however, if the bias resulting from a particular estimation procedure is independent of the true value of the underlying parameter. In generalized linear models this is not the case and we need to introduce a modification.

We shall illustrate the procedure with a simple 2-level variance components model, as follows

$$\text{logit}(\pi_{ij}) = \beta_0 + \beta_1 x_{ij} + u_j$$

$$u_j \sim N(0, \sigma_u^2)$$

$$y_{ij} \sim Binomial(1, \pi_{ij})$$

Given a set of initial estimates, obtained using for example the first order MQL approximation,

$$\hat{\sigma}_u^{2(0)}, \quad \hat{\beta}_0^{(0)}, \quad \hat{\beta}_1^{(0)} \tag{4.4.1}$$

we generate a set of bootstrap samples, parametrically or using the residuals bootstrap (see section 3.6) from the model using the estimates (4.4.1) and averaging over these we obtain the set of bootstrap estimates

$$\tilde{\sigma}_u^{2(0)}, \quad \tilde{\beta}_0^{(0)}, \quad \tilde{\beta}_1^{(0)} \tag{4.4.2}$$

We now obtain the bootstrap estimate of the bias by subtracting (4.4.2) from (4.4.1). These bias estimates are added to the initial parameter estimates (4.4.1) as a first adjustment to give new bias-corrected estimates

$$\hat{\sigma}_u^{2(1)}, \quad \hat{\beta}_0^{(1)}, \quad \hat{\beta}_1^{(1)} \tag{4.4.3}$$

We generate a new set of bootstrap samples from the model based upon the estimates given by (4.4.3), subtract the new mean bootstrap parameter estimates from (4.4.3) to obtain updated bias estimates and add these to the initial estimates (4.4.1) to obtain a new set of bias-corrected estimates. When it converges, Kuk (1995) demonstrates that this procedure gives asymptotically consistent and unbiased parameter estimates.

Care needs to be taken with small variance estimates. To estimate the bias we need to allow negative estimates of variances. If an initial estimate is zero, then clearly, resetting negative bootstrap sample means to zero implies that the bias estimate will never be negative, so the new updated estimate will remain at zero. Moreover, as confirmed by simulations, all the estimates will exhibit a downward bias if negative bootstrap means are reset to zero. We also note that where an unbiased variance estimate is close to zero, the value of the bias is anyway small, so that full bias correction is less important and, for example, a second order PQL estimate may be adequate.

The bootstrap replicates from the final bootstrap set generally will have too small a variance and so cannot directly be used for inference. If we knew the functional relationship between the bias-corrected value and the biased value this could be used to transform each of the bootstrap replicate estimates and the transformed values then used for inference. Alternatively, for each parameter in turn, using the final bias-corrected estimate and the final bootstrap replicate mean, we take the ratio of these and multiply all the final replicate parameter values by this ratio. These scaled values can be used to construct approximately correct standard errors and quantile estimates. Care is needed, however, when the initial parameter estimates are close to zero.

5

Models for Repeated Measures Data

5.1 Repeated measures data

When measurements are repeated on the same subjects, for example students or animals, a 2-level hierarchy is established with measurement repetitions or occasions as level 1 units and subjects as level 2 units. Such data are often referred to as 'longitudinal' as opposed to 'cross-sectional' where each subject is measured only once. Thus, we may have repeated measures of body weight on growing animals or children, repeated test scores on students or repeated interviews with survey respondents. It is important to distinguish two classes of models which use repeated measurements on the same subjects. In one, earlier measurements are treated as covariates rather than responses. This was done for the educational data analysed in Chapters 2 and 3, and usually will be appropriate when there are a small number of discrete occasions and where different measures are used at each one. In the other, usually referred to as 'repeated measures' models, all the measurements are treated as responses, and it is this class of models we shall discuss here. A detailed description of the distinction between the former 'conditional' models and the latter 'unconditional' models can be found in Goldstein (1979) and Plewis (1985). Singer and Willett (2002) give detailed examples of applications both of repeated measures and conditional models.

We may also have repetition at higher levels of a data hierarchy. For example, we may have annual examination data on successive cohorts of 16 year-old students in a sample of schools. In this case the school is the level 3 unit, year is the level 2 unit and student the level 1 unit. We shall also look at an example where there are responses at both level 1 and level 2, that is specific to the occasion and to the subject. It is worth pointing out that in repeated measures models typically most of the variation is at level 2, so that the proper specification of a multilevel model for the data is of particular importance.

The link with the multivariate data models in Chapter 6 is also apparent when the occasions are fixed. For example, we may have measurements on the height of a sample of children at ages 11.0, 12.0, 13.0 and 14.0 years. We can regard this as consisting of a multivariate response vector of four responses for each child, and perform an equivalent analysis, for example relating the measurements to a polynomial function of age. This multivariate approach has traditionally been used with repeated measures data (see Grizzle and Allen, 1969). It cannot, however, deal properly with data with an arbitrary spacing or number of occasions and need not be considered further.

In all the models considered so far we have assumed that the level 1 residuals are uncorrelated. For some kinds of repeated measures data, however, this assumption will not be reasonable, and we shall investigate models which allow a serial correlation structure for these residuals.

We deal first with continuous response variables and we shall discuss repeated measures models for discrete response data later.

5.2   A 2-level repeated measures model

Consider a dataset consisting of repeated measurements of the heights of a random sample of children. We can write a simple model

$$y_{ij} = \beta_{0j} + \beta_{1j}x_{ij} + e_{ij} \tag{5.1}$$

This model assumes that height (Y) is linearly related to age (X) with each subject having their own intercept and slope so that

$$E(\beta_{0j}) = \beta_0, \quad E(\beta_{1j}) = \beta_1$$

$$\text{var}(\beta_{0j}) = \sigma_{u0}^2, \quad \text{var}(\beta_{1j}) = \sigma_{u1}^2, \quad \text{cov}(\beta_{0j}, \beta_{1j}) = \sigma_{u01}, \quad \text{var}(e_{ij}) = \sigma_e^2$$

There is no restriction on the number or spacing of ages, so that we can fit a single model to subjects who may have one or several measurements. We can clearly extend (5.1) to include further explanatory variables, measured either at the occasion level, such as time of year or state of health, or at the subject level, such as birthweight or gender. We can also extend the basic linear function in (5.1) to include higher order terms and we can further model the level 1 residual so that, for example, the level 1 variance is a function of age.

We shall explore briefly a nonlinear model for growth measurements in Chapter 8. Such models have an important role in certain kinds of growth modelling, especially where growth approaches an asymptote as in the approach to adult status in animals. In the following sections we shall discuss the use of polynomial models which have a more general applicability and for many applications are more flexible (for a further discussion see Goldstein 1979). We introduce examples of increasing complexity.

5.3   A polynomial model example for adolescent growth and the prediction of adult height

Our first example combines the basic 2-level repeated measures model with a multi-variate model to show how a general growth prediction model can be constructed. The data consist of 436 measurements of the heights of 110 boys between the ages of 11 and 16 years together with measurements of their height as adults and estimates of their bone ages at each height measurement, based upon wrist radiographs. A detailed description can be found in Goldstein (1989b). We first write down the three basic components of the model, starting with a simple repeated measures model for height using a 5-th degree polynomial:

$$y_{ij}^{(1)} = \sum_{h=0}^{5} \beta_h^{(1)} x_{ij}^h + \sum_{h=0}^{2} u_{hj}^{(1)} x_{ij}^h + e_{ij}^{(1)} \tag{5.2}$$

where the level 1 term e_{ij} may have a complex structure, for example a decreasing variance with increasing age.

The measure of bone age is already standardized since the average bone age for boys of a given chronological age is equal to this age for the population. Thus we model bone age using an overall constant to detect any average departure for this group together with between-individual and within-individual variation.

$$y_{ij}^{(2)} = \beta_0^{(2)} + \sum_{h=0}^{1} u_{hj}^{(2)} x_{ij}^h + e_{ij}^{(2)} \tag{5.3}$$

For adult height we have a simple model with an overall mean and level 2 variation. If we had more than one adult measurement on individuals we would be able to estimate also the level 1 variation among adult height measurements; in effect measurement errors.

$$y_j^{(3)} = \beta_0^{(3)} + u_{0j}^{(3)} \tag{5.4}$$

We now combine these into a single model using the following indicators:

$$\delta_{ij}^{(1)} = 1, \quad \text{if growth period measurement,} \quad 0 \;\; \text{otherwise}$$
$$\delta_{ij}^{(2)} = 1, \quad \text{if bone age measurement,} \quad 0 \;\; \text{otherwise}$$
$$\delta_j^{(3)} = 1, \quad \text{if adult height measurement,} \quad 0 \;\; \text{otherwise}$$

$$y_{ij} = \delta_{ij}^{(1)} \left(\sum_{h=0}^{5} \beta_h^{(1)} x_{ij}^h + \sum_{h=0}^{2} u_{hj}^{(1)} x_{ij}^h + e_{ij}^{(1)} \right)$$
$$+ \delta_{ij}^{(2)} \left(\beta_0^{(2)} + \sum_{h=0}^{1} u_{hj}^{(2)} x_{ij}^h + e_{ij}^{(2)} \right) + \delta_j^{(3)} \left(\beta_0^{(3)} + u_{0j}^{(3)} \right) \tag{5.5}$$

At level 1 the simplest model, which we shall assume, is that the residuals for bone age and height are independent, although dependencies could be created, for example if the model was incorrectly specified at level 2. Thus, level 1 variation is specified in terms of two variance terms. Although the model is strictly a multivariate model, because the level 1 random variables are independent, it is unnecessary to specify a 'dummy' level 1 with no random variation as with standard multivariate models (see Chapter 6). If, however, we allow correlation between height and bone age then we will need to specify the model with no variation at level 1, the variances and covariance between bone age and height at level 2 and the between-individual variation at level 3.

Table 5.1 shows the fixed and random parameters for this model, omitting the estimates for the between-individual variation in the quadratic coefficient of the polynomial growth curve. We see that there is a large non-zero correlation between adult height and height and small correlations between adult height and the height growth and the bone age coefficients. This implies that the height and bone age measurements can be used to make predictions of adult height. In fact these predicted values are simply the estimated residuals for adult height. For a new individual, with information available at one or more ages on height or bone age, we simply estimate the adult height residual using the model parameters. Table 5.2 shows the estimated standard errors associated with predictions made on the basis of varying amounts of information. It is clear that

Table 5.1 Height (cm) for adolescent growth, bone age, and adult height for a sample of boys. Age measured about 13.0 years. Level 2 variances and covariances shown; correlations are in parentheses

(a) Fixed

Parameter	Estimate (s.e.)
Adult height	
Intercept	174.4
Group (A–B)	0.25 (0.50)
Height	
Intercept	153.0
Age	6.91 (0.20)
Age^2	0.43 (0.09)
Age^3	−0.14 (0.03)
Age^4	−0.03 (0.01)
Age^5	0.03 (0.03)
Bone age	
Intercept	0.21 (0.09)

(b) Random

	Adult height	Height intercept	Age	Bone age intercept
Level 2 variances				
Adult height	62.5			
Height intercept	49.5 (0.85)	54.5		
Age	1.11 (0.09)	1.14 (0.09)	2.5	
Bone age intercept				0.85
Level 1 variances				
Height	0.89			
Bone age	0.18			

Table 5.2 Standard errors for adult height predictions for specified combinations of height and bone age measurements

	Height measures (age)		
	None	11.0	12.0 / 8.0
Bone age measures			
None		4.3	4.2
11.0	7.9	3.9	3.8
12.0	7.9	3.7	3.7

the main gain in efficiency comes with the use of height with a smaller gain from the addition of bone age.

The method can be used for any measurements, either to be predicted or as predictors. In particular, covariates such as family size or social background can be included to improve the prediction. We can also predict other events of interest, such as the estimated age at maximum growth velocity.

Pan and Goldstein (1997) derive a procedure based on such a model for estimating norms for complex growth functions of height and weight such as acceleration coefficients at particular ages from whatever combination of measurements happens to be available. After initial Normalization of the data they use a longitudinal standardizing sample to establish population estimates of the growth curve polynomial coefficients covariance matrix. This is then used to derive the distribution of the required growth functions.

Pan and Goldstein (1998) extend the basic polynomial model by considering spline functions that are smoothly joining polynomials with fixed join points or 'knots'. Thus, for a set of measurements on head circumference of children from birth to 16 years, after some exploration they consider the following model:

$$y_{ij} = \beta_{0j} + \beta_{1j}t_{ij} + \beta_{2j}t_{ij}^2 + \beta_{3j}\log(12t_{ij} + 1) + \beta_{4j}(\xi - t_{ij})_+^3 + \beta_{5j}(t_{ij} - \xi)_+^3 + e_{ij}$$

$$\theta_+ = \begin{cases} \theta & \text{if } \theta > 0 \\ 0 & \text{if } \theta \leq 0 \end{cases}$$

with knots at 2.0 and 10.0 years. The advantage of such models is that, while there is an underlying polynomial across the whole age range, the local end relationships are further modelled by the '+' function components. This, at least in part, overcomes a disadvantage of ordinary polynomials that typically fail to provide good fits at the extremes of the time period. Note that the first + function is present up to the age of 2.0 years and the second from 10.0 years onwards. Another example of the use of such splines is given by Blatchford et al. (2002) in modelling educational achievement as a function of class size.

5.4 Modelling an autocorrelation structure at level 1

So far we have assumed that the level 1 residuals are mutually independent. In many situations, however, such an assumption would be false. For growth measurements the specification of level 2 variation serves to model a separate curve for each individual, but the between-individual variation will typically involve only a few parameters, as in the previous example. Thus if measurements on an individual are obtained very close together in time, they will tend to have similar departures from that individual's underlying growth curve. That is, there will be 'autocorrelation' between the level 1 residuals. Examples arise from other areas, such as economics, where measurements on each unit, for example an enterprise or economic system, exhibit an autocorrelation structure and where the parameters of the separate time series vary across units at level 2.

A detailed discussion of multilevel time series models is given by Goldstein et al. (1994). They discuss both the discrete time case, where the measurements are made at the same set of equal intervals for all level 2 units, and the continuous time case where the time intervals can vary. We shall develop the continuous time model here since it is both more general and flexible.

To simplify the presentation, we shall drop the level 1 and 2 subscripts and write a general model for the level 1 residuals as follows:

$$\text{cov}(e_t e_{t-s}) = \sigma_e^2 f(s) \tag{5.6}$$

Thus, the covariance between two measurements depends on the variance and the time difference between the measurements. The function $f(s)$ is conveniently described by

Table 5.3 Some choices for the covariance function g for level 1 residuals

$g = \beta_0 s$	For equal intervals this is a first order autoregressive series
$g = \beta_0 s + \beta_1(t_1 + t_2) + \beta_2(t_1^2 + t_2^2)$	For time points t_1, t_2 this implies that the variance is a quadratic function of time
$g = \begin{cases} \beta_0 s & \text{if no replicate} \\ \beta_1 & \text{if replicate} \end{cases}$	For replicated measurements this gives an estimate of measurement reliability $\exp(-\beta_1)$
$g = (\beta_0 + \beta_1 z_{1j} + \beta_2 z_{2ij})s$	The covariance is allowed to depend on an individual-level characteristic (e.g. gender) and a time-varying characteristic (e.g. season of the year or age)
$g = \begin{cases} \beta_0 s + \beta_1 s^{-1}, & s > 0 \\ 0, & s = 0 \end{cases}$	Allows a flexible functional form, where the time intervals are not close to zero

a negative exponential reflecting the common assumption that with increasing time difference the covariance tends to a fixed value, $\alpha\sigma_e^2$, and typically this is assumed to be zero

$$f(s) = \alpha + \exp(-g(\beta, z, s)) \tag{5.7}$$

where β is a vector of parameters for explanatory variables z. Some choices for g are given in Table 5.3.

We can apply the methods described in Appendix 8.1 to obtain maximum likelihood estimates for these models, by writing the expansion

$$f(s, \beta, z) = \left\{1 + \sum_k \beta_{k,t} z_k g(H_t)\right\} f(H_t) - \sum_k \beta_{k,t+1} z_k g(H_t) f(H_t) \tag{5.8}$$

so that the model for the random parameters is linear. Full details are given by Goldstein et al. (1994). Pourahmadi (1999, 2000) considers similar models but restricted to a fixed set of discrete occasions and using a different algorithm to obtain maximum likelihood estimates.

5.5 A growth model with autocorrelated residuals

The data for this example consist of a sample of 26 boys each measured on nine occasions between the ages of 11 and 14 years (Harrison and Brush, 1990). The measurements were taken approximately 3 months apart. Table 5.4 shows the estimates from a model which assumes independent level 1 residuals with a constant variance. The model also includes a cosine term to model the seasonal variation in growth with time measured from the beginning of the year. If the seasonal component has amplitude α and phase γ we can write

$$\alpha \cos(t + \gamma) = \alpha_1 \cos(t) - \alpha_2 \sin(t)$$

In the present case the second coefficient is estimated to be very close to zero and is set to zero in the following model. This component results in an average growth difference between summer and winter estimated to be about 0.5 cm.

We now fit in Table 5.5 the model with $g = \beta_0 s$ which is the continuous time version of the first order autoregressive model.

Table 5.4 Height as a fourth degree polynomial on age, measured about 13.0 years. Standard errors in parentheses; correlations in parentheses for covariance terms

(a) Fixed

Parameter	Estimate (s.e.)
Intercept	148.9
Age	6.19 (0.35)
Age^2	2.17 (0.46)
Age^3	0.39 (0.16)
Age^4	−1.55 (0.44)
cos(time)	−0.24 (0.07)

(b) Random

	Intercept	Age	Age^2
Level 2 covariance matrix			
Intercept	61.6 (17.1)		
Age	8.0 (0.61)	2.8 (0.7)	
Age^2	1.4 (0.22)	0.9 (0.67)	0.7 (0.2)
Level 1 variance			
σ_e^2	0.20 (0.02)		

Table 5.5 Height as a fourth degree polynomial on age, measured about 13.0 years. Standard errors in parentheses; correlations in parentheses for covariance terms. Autocorrelation structure fitted for level 1 residuals

(a) Fixed

Parameter	Estimate (s.e.)
Intercept	148.9
Age	6.19 (0.35)
Age^2	2.16 (0.45)
Age^3	0.39 (0.17)
Age^4	−1.55 (0.43)
cos (time)	−0.24 (0.07)

(b) Random

	Intercept	Age	Age^2
Level 2 covariance matrix			
Intercept	61.5 (17.1)		
Age	7.9 (0.61)	2.7 (0.7)	
Age^2	1.5 (0.25)	0.9 (0.68)	0.6 (0.2)
Level 1 parameters			
σ_e^2	0.23 (0.04)		
β	6.90 (2.07)		

The fixed part and level 2 estimates are little changed. The autocorrelation parameter implies that the correlation between residuals 3 months (0.25 years) apart is 0.19.

Finally, on this topic, there will typically need to be a trade-off between modelling more random coefficients at level 2 in order to simplify or eliminate a level 1 serial correlation structure, and modelling level 2 in a parsimonious fashion so that a relatively small number of random coefficients can be used to summarize each individual. An extreme example of the latter is given by Diggle (1988), who fits only a random intercept at level 2 and serial correlation at level 1.

5.6   Multivariate repeated measures models

We have already discussed the bivariate repeated measures model where the level 1 residuals for the two responses are independent. In the general multivariate case where correlations at level 1 are allowed, we can fit a full multivariate model by adding a further lowest level as described in Chapter 6. For the autocorrelation model this will involve extending the models to include cross-correlations. For example for two response variables with the model of Table 5.5 we would write

$$g = \sigma_{e1} \sigma_{e2} \exp(-\beta_{12} s)$$

The special case of a repeated measures model where some or all occasions are fixed is of interest. We have already dealt with one example of this where adult height is treated separately from the other growth measurements. The same approach could be used with, for example, birthweight or length at birth. In some studies all individuals may be measured at the same initial occasion and we can choose to treat this as a covariate rather than as a response. This might be appropriate where individuals were divided into groups for different treatments following initial measurements.

Having fitted a multivariate model this allows us to derive the correlation matrix between measurements for given time differences. Thus, for example, if we have a bivariate model with height and weight as responses we can compare the correlation as a function of time interval between earlier weight and later height with the correlation between earlier height and later weight. If the latter correlation, for a given time interval, was substantially greater than the former we might (tentatively) infer that earlier height influenced subsequent weight, in a causal sense, rather than the other way around.

5.7   Scaling across time

For some kinds of data, for example educational achievement scores, different measurements may be taken over time on the same individuals so that some form of standardization may be needed before they can be modelled using the methods of this chapter. It is common in such cases to standardize the measurements so that at each measuring occasion they have the same population distribution. If this is done then we should not expect any trend in either the mean or variance over time, although there will still, in general, be between-individual variation. An alternative standardization procedure is to convert scores to age equivalents; that is to assign to each score the age for which that score is the population mean or median. Where scores change smoothly with age this has the attraction of providing a readily interpretable scale.

Plewis (1993) uses a variant of this in which the coefficient of variation at each age is also fixed to a constant value. In general, different standardizations may be expected to lead to different inferences. The choice of standardization is in effect a choice about the appropriate scale along which measurements can be equated so that any interpretation needs to recognize this. A further discussion of this issue is given by Plewis (1996).

We could also fit unstandardized means at each time point, or as a smoothly varying function of time. Similarly, rather than standardize the variance prior to modelling we can choose to model it also as a function of time (see Chapter 3), and this will be important when the measurements are not made at a small number of discrete occasions.

5.8 Cross-over designs

A common procedure for comparing the effects of two different treatments A, B, is to divide the sample of subjects randomly into two groups and then to assign A to one group followed by B, and B to the other group followed by A. The potential advantage of such a design is that the between-individual variation can be removed from the treatment comparison. A basic model for such a design with two treatments, repeated measurements on individuals and a single group effect can be written as follows:

$$y_{ij} = \beta_0 + \beta_1 x_{1ij} + \beta_2 x_{2ij} + u_{0j} + u_{2j} x_{2ij} + e_{ij} \tag{5.9}$$

where X_1 is a dummy variable for time period and X_2 is a dummy variable for treatment. In this model we have not modelled the responses as a function of time within treatment, but this can be added in the standard fashion described in previous sections. In the random part at level 2 we allow between-individual variation for the treatment difference and we can also structure the level 1 variance to include autocorrelation or different variances for each treatment or time period.

One of the problems with such designs is so-called 'carry-over' effects whereby exposure to an initial treatment leaves some individuals more or less likely to respond positively to the second treatment. In other words, the u_{2j} may depend on the order in which the treatments were applied. To model this we can add an additional term to the random part of the model, say $u_{3j} \delta_{3ij}$, where δ_{3ij} is a dummy variable which is 1 when A precedes B and the second treatment is being applied and zero otherwise. This will also have the effect of allowing level 2 variances to depend on the ordering of treatments. Putt and Chinchilli (1999) further extend the modelling of carry-over effects.

5.9 Missing data

In repeated measures studies that are designed to follow up samples of individuals it is often the case that measurements are not made at one or more target times or occasions. We have already shown that 'balanced' data are not a requirement for efficient estimates and some studies, for example, 'rotate' individuals or higher level units in or out of the study by design. Where, however, missing measurements on individuals or units are unscheduled, certain important issues arise. Three situations can be distinguished. Laird (1988) provides a detailed discussion.

The first case is where missingness is completely at random (MCAR), that is the probability of being missing, the 'non-response' probability, is independent of any of the other responses for that individual or higher level unit. This is effectively equivalent to the 'missing by design' case and all the procedures we have discussed can be applied.

The second case is where missingness is at random (MAR), that is the non-response probability depends only on observed responses. Thus, for example in a health interview study, previous ill health may increase the probability of non-attendance at a subsequent interview. We may also include here the possibility that we can achieve MAR by introducing covariates into the model. Thus, for example, social background may affect the probability of non-response over time. Broadly, so long as the model specifying the covariance structure of the responses is correct, we can apply our previous procedures. The joint probability of the observed responses, given the model parameters, is independent of the joint probability of the unobserved responses given the parameters, so that any estimation method, such as maximum likelihood or MCMC which is based upon the joint probability function of the observations will provide valid inferences.

The third case is more troublesome, and it is where the probability of reponse is not independent of the unobserved, missing, values. In general, our previous procedures will lead to biases unless they are suitably modified. One approach to the problem, the 'pattern mixture' model, essentially groups individuals according to different patterns of missing responses. The patterns can have different models for the non-response, so that the response values are considered as dependent on the response probabilities. Another approach, which seems more natural, is to suppose that the response probabilities are dependent on the data, both observed and unobserved, and also are a function of time. These models are known as selection models. Diggle and Kenward (1994) develop such a model for 'monotone drop out' where individuals do not return to a study once they fail to respond at a target occasion. Kenward (1998) further develops this and introduces a sensitivity analysis to study how robust inferences are to choice of model structure. In these cases a model for the response missingness mechanism is modelled jointly with the target model, for example a polynomial growth curve. Since the model for the probability of response depends on unknown, that is missing, response values, estimation procedures such as EM or MCMC that can deal with 'imputing' missing values can be used.

Touloumi et al. (1999) gives an example of such a model, using EM estimation, where changing cell counts, in a growth component, is modelled jointly with (log) patient survival time and where dropout is not independent of the survival time. In their model they do not have an explicit model for the probability of dropout, but essentially impute the unknown survival time at each iteration of the algorithm. Crouchley and Ganjali (2002) provide a general treatment of these models.

A useful specification for these models is to employ the probit link function in the response model as described in Appendix 4.3, assuming multivariate Normality for the growth component. This then estimates the covariance at the occasion level between the 'propensity to respond' and the growth measurement. This covariance can be a function of time or time-related covariates and such models can, in principle, be fitted within an MCMC framework and are not restricted to monotone dropout structures. At the level of the individual we allow the response to incorporate random effects which will covary with the individual random effects for the growth model, although in the monotone dropout case we will generally not be able to fit separate random effects at the occasion and individual level. The model generalizes straightforwardly, using the procedures we have outlined, to multivariate growth processes.

5.10   Longitudinal discrete response data

If we have repeated measures for discrete responses then a natural approach is to use the models of Chapter 4 in a 2-level model with occasions as level 1. Thus we could write for a binary response

$$f(\pi_{ij}) = \beta_{0j} + \beta_{1j}x_{ij}$$

$$E(\beta_{0j}) = \beta_0, \quad E(\beta_{1j}) = \beta_1$$

$$\text{var}(\beta_{0j}) = \sigma_{u0}^2, \quad \text{var}(\beta_{1j}) = \sigma_{u1}^2, \quad \text{cov}(\beta_{0j}, \beta_{1j}) = \sigma_{u01}, \quad \text{var}(e_{ij}) = \sigma_e^2 \qquad (5.10)$$

$$y_{ij} \overset{iid}{\sim} \text{Bin}(1, \pi_{ij})$$

with straightforward extensions to ordered and unordered responses. In some cases such a formulation may be reasonable but often the assumption that the binomial responses are independent and identically distributed (*iid*) conditional on the π_{ij}, is unrealistic. Thus, for example, in a study of voting patterns on three occasions Yang et al. (2000) used a 3-level model for repeated binary responses of voting behaviour. They found that there was considerable under-dispersion in the data with an extra-binomial parameter as low as 0.4 and this was attributed to the fact that for many people the probability of voting for a particular party (Conservative) was effectively 0 or 1, which implies in (5.10) large numbers of individuals with infinite random effect values on the linear scale.

One method for handling such data is the so-called mover-stayer model which extends (5.10) by writing the probability of being a 'stayer', that is not changing, as

$$\pi_{1j} = f[(X_1\alpha_1)_j + u_{1j}]$$

the probability of voting Conservative given that the individual is a stayer as

$$\pi_{2ij} = f[(X_2\alpha_2)_{ij} + u_{2j}]$$

and the probability of voting Conservative if the individual is a mover as

$$\pi_{3ij} = f[(X_3\alpha_3)_{ij} + u_{3j}]$$

For those individuals who have non-constant responses the probability of being a stayer is 0. We can combine these probabilities with (5.10) so that the probability of voting Conservative is

$$\pi_{1j}\pi_{2ij} + (1 - \pi_{1j})\pi_{3ij}$$

and the relevant parameters can be estimated, for example by maximum likelihood or MCMC.

One problem with this formulation is the assumption that people can be stayers and that this describes the response pattern. In the voting example with only three occasions this seems questionable when nearly three quarters of the sample did not change their vote. In other cases, for example in modelling the presence of a disease over time, such an assumption may also be questionable.

To avoid relying on such a model we can directly model the responses as multivariate and estimate the covariance structure, where occasions are treated as variates. Thus in the voting case we would have a three-variate model which we can fit as an eight-category multinomial response (see Chapter 4) with three dummy variable terms for occasion representing the average response at each occasion. We can also include covariates, possibly interacting with these dummy variables. At each occasion the coefficient of each dummy variable is assumed to vary binomially across individuals and the covariances between these are parameters to be estimated. A simple intercept model can be written

$$\text{logit}(\pi_{tij}) = \sum_{t=1}^{3} \beta_t z_{tij}, \quad y_{tij} \sim \text{Bin}(1, \pi_{tij}) \tag{5.11}$$

where t indexes occasion and the z_{tij} are the dummy variables for occasions. If we use a probit link function we can directly estimate and interpret these covariances as correlations.

This multivariate formulation, however, as in the case of continuous responses, is relatively inflexible and Barbosa and Goldstein (2000) extend this model to the general case with arbitrary numbers and spacing of occasions. Essentially they use the serial correlation models in section 5.5 adapted for binary responses where the correlation between occasions becomes a function of the time difference. They find that the final choice of covariance function in Table 5.3 involving an inverse polynomial fits the data well and show that this is an efficient method for handling such longitudinal data and accounting for the under-dispersion that results from a large proportion of unchanging responses. The method is readily generalized to the ordered or unordered category case, although the latter will generally lead to a large number of potential parameters. A particular advantage of this method is that it models the correlation between occasions as a smooth function of time, and possibly covariates, so allowing flexible predictions of future probabilities based on current observations to be made for individuals or groups of individuals.

6

Multivariate Multilevel Data

6.1 Introduction

In previous chapters we largely considered only a single response variable, although we looked at special cases of multiple responses such as in describing repeated measures data. We now look more systematically at models where we wish simultaneously to model several responses as functions of explanatory variables. We confine ourselves to the case of continuous responses; for the discrete response case the methods of Chapter 4 are readily adapted. In particular, as pointed out in Chapter 4, where a probit link function is used we can provide estimates for parameters which are all derived from an underlying multivariate Normal structure.

The formulation of multivariate models given here provides us with tools for tackling a very wide range of problems. These problems include missing response data and rotation or matrix designs for surveys. We develop the model using a dataset of examination results.

The data used as an example consist of scores on two components of a science examination taken in 1989 by 1905 students in 73 schools and colleges. The examination is the General Certificate of Secondary Education (GCSE) taken at the end of compulsory schooling in England, normally when students are 16 years of age. The first component is a traditional written question paper (marked out of a total score of 160) and the second consists of coursework (marked out of a total score of 108), including projects undertaken during the course and marked by each student's own teacher. The overall teachers' marks are subject to external 'moderation' using a sample of coursework. Interest in these data centres on the relationship between the component marks at both the school and student level, whether there are gender differences in this relationship and whether the variability differs for the two components. Creswell (1991) has a full description of the dataset.

6.2 The basic 2-level multivariate model

To define a multivariate, in the case of our example a 2-variate, model we treat the individual student as a level 2 unit and the 'within-student' measurements as level 1 units. Each level 1 measurement 'record' has a response, which is either the written paper score or the coursework score. The basic explanatory variables are a set of dummy variables that indicate which response variable is present. Further explanatory variables

Table 6.1 Data matrix for examination example

Student	Response	Intercepts		Gender	
		Written	Coursework	Written	Coursework
1 (female)	y_{11}	1	0	1	0
1	y_{21}	0	1	0	1
2 (male)	y_{12}	1	0	0	0
2	y_{22}	0	1	0	0
3 (female)	y_{13}	1	0	1	0

are defined by multiplying these dummy variables by individual-level explanatory variables, for example gender.

The data matrix for three individuals, two of whom have both measurements and the third of whom has only the written paper score, is displayed in Table 6.1. The first and third students are female (1) and the second is male (0).

The model is written as

$$y_{ij} = \beta_{01}z_{1ij} + \beta_{02}z_{2ij} + \beta_{11}z_{1ij}x_j + \beta_{12}z_{2ij}x_j + u_{1j}z_{1ij} + u_{2j}z_{2ij}$$

$$z_{1ij} = \begin{Bmatrix} 1 & \text{if written} \\ 0 & \text{if coursework} \end{Bmatrix}, \quad z_{2ij} = 1 - z_{1ij}, \quad x_j = \begin{Bmatrix} 1 & \text{if female} \\ 0 & \text{if male} \end{Bmatrix} \quad (6.1)$$

$$\text{var}(u_{1j}) = \sigma_{u1}^2, \quad \text{var}(u_{2j}) = \sigma_{u2}^2, \quad \text{cov}(u_{1j}u_{2j}) = \sigma_{u12}$$

There are several features of this model. There is no level 1 variation specified because level 1 exists solely to define the multivariate structure. The level 2 variances and covariance are the (residual) between-student variances. In the case where only the intercept dummy variables are fitted, and since every student has both scores, the model estimates of these parameters become the usual between-student estimates of the variances and covariance. The multilevel estimates are statistically efficient even where some responses are missing, and in the case where the measurements have a multivariate Normal distribution they are maximum likelihood. Thus the formulation as a 2-level model allows for the efficient estimation of a covariance matrix with missing responses.

In our example the students are grouped within examination centres, so that the centre is the level 3 unit. Table 6.2 presents the results of two models fitted to these data. The first analysis is simply (6.1) with variances and a covariance for the two components added at level 3. In the second analysis additional variance terms for gender have been added at levels 2 and 3.

In both analyses the females do worse on the written paper and better on the coursework assessment. There is a greater variability of marks on the coursework element, even though this is marked out of a smaller total, and the centre variance partition coefficients are approximately the same in the first analysis (0.28 and 0.30). This suggests that the 'moderation' process has been successful in maintaining a similar relative between-centre variation for the coursework marks. The correlation between the two components is 0.50 at the student level and 0.41 at the centre level.

In the second analysis we see that the between-student variance for coursework is smaller for the females (164.0) compared to that for the males (189.1) and for the centres the coursework variance for females is also smaller (73.3) than for males (106.6). There

Table 6.2 Bivariate models for written paper and coursework responses

	Estimate (s.e.)	Estimate (s.e.)
Fixed		
Constant		
Written	49.5	49.5
Coursework	69.5	60.1
Female		
Written	−2.5 (0.5)	−2.5 (0.5)
Coursework	6.9 (0.7)	7.3 (1.1)
Random		
Level 3		
σ_{v1}^2	48.9 (9.5)	49.6 (9.5)
σ_{v12}	25.2 (9.1)	35.5 (11.3)
σ_{v2}^2	77.1 (14.8)	106.6 (21.7)
σ_{v14}		−15.9 (7.8)
σ_{v24}		−37.4 (13.2)
σ_{v4}^2		41.5 (11.7)
Level 2		
σ_{u1}^2	124.3 (4.1)	124.2 (4.1)
σ_{u12}	74.6 (3.9)	73.6 (3.9)
σ_{u2}^2	183.2 (6.1)	189.1 (8.6)
σ_{u24}		−12.5 (4.7)
−2 (log-likelihood)	29718.8	29664.7

For the random parameters the subscripts refer to the following explanatory variables: 1 = writing intercept; 2 = coursework intercept; 3 = writing female; 4 = coursework female.

appears to be no difference in the variances for the written paper and the corresponding parameter is omitted.

Note how the standard error of the coursework gender coefficient increases with the more precise specification of the coursework variation at both levels. This is another aspect of the effect we saw when fitting a multilevel model as opposed to a single-level model.

6.3 Rotation designs

We have already seen that fully balanced multivariate designs are unnecessary when we formulate the model as in Table 6.1. As this shows, the basic 2-level formulation does not formally recognize that a response is missing, since we only record those present.[1] We now look at designs where responses are deliberately missing by design and we see how this can be useful in a number of circumstances.

In many kinds of surveys the amount of information required from respondents is so large that it is too onerous to expect each one to respond to all the questions or items. In education we may require achievement information covering a large number of areas, in surveys of businesses we may wish to have a large amount of detailed information from each business, and in household questionnaires we may wish to obtain information

---

[1] In MCMC estimation it is convenient to estimate the 'missing' responses at each iteration so that subsequent steps can condition on a balanced data set.

on a wide range of topics. We consider only measurements that are used as responses in a model. If we denote the total set of responses as $\{N\}$ then we choose p subsets $\{N_i, i = 1, \ldots p\}$ each of which is suitable for administering to a level 1 unit such as a student or household.

When choosing these subsets we can only estimate subject-level covariances between those responses that appear together in a subtest. It is therefore common in such designs to ensure that every possible pair of responses is present. If we wish to estimate covariances for higher level units such as schools it is necessary only to ensure that the relevant pair of responses are assigned to some schools – a large enough number to provide efficient estimates. The subjects are assigned at random to subtest and higher level units are also assigned randomly, possibly with stratification.

Each subset is viewed formally as a multivariate response vector with randomly missing values, that is those that are excluded from the subset. As we saw in the previous section, we can fit a multivariate response model for such data and obtain efficient estimates for the fixed part coefficients and covariance structures at any level. In this formulation, the variables to be used as explanatory variables should be measured for each level 1 unit. We give an example using educational achievement data. In Chapter 7 we shall look at examples where a small number of underlying factors is used to account for the observed covariance structures.

6.4 A rotation design example using science test scores

The data come from the Second International Science Survey carried out by the International Association for the Evaluation of Educational Achievement (Rosier, 1987). Table 6.3 shows how items from three science topic areas are distributed over test papers or forms with the numbers of items in each topic area. The tests consisted of a core form, or subtest, taken by all students plus a randomly selected pair out of the four additional forms or subtests. The study was carried out in 1984 in some 24 countries. We discuss here the results for Hungary.

Because the number of items in the first additional form was very small, and likewise in some of the other forms for some subjects, only the additional forms 2–4 are used, labelled Biology R3, Biology R4 and Physics R2 (Table 6.4). We also divide each subtest score by the total number of items in the subtest so as to reduce each score to the same scale. There are 99 schools with 2439 students and a total of 10 971 responses.

We see that the intercorrelations at the student level are low and are higher at the school level. One reason for this is the fact that there are few items in each subtest so that the reliability of the tests is rather low. This will decrease the correlations at the student level but less so at the school level. Because of these low correlations among the subtests the joint analysis does not result in a marked improvement in efficiency

Table 6.3 Numbers of items in topic areas: Grade 8

Form	Earth Science	Biology	Physics
1(Core)	6	10	10
2	–	–	7
3	–	4	–
4	–	4	–

when we compare this analysis with an analysis for a single subtest. For example, if we fit a univariate model for the Physics R2 subtest, using the 1226 students responding to that subtest, we obtain fixed part estimates of 0.665 (0.0132) and −0.073 (0.0124) which are close to those above and with standard errors only slightly higher.

In order to provide the most precise estimates we treated the subtests separately, although we would generally wish to make inferences for each subject area, combining over the tests. The natural way to do this is to form a weighted average of the subtest estimates, in this case weighting by the number of items in each subtest. Thus, for the biology core and subtests we would form the weighted sum with weights 0.556, 0.222 and 0.222 respectively. This gives estimates for the boys and (girls–boys) of 0.68 (0.009) and −0.02 (0.007).

We can compare this with simply using the original scores and forming the weighted combination of the core and two subtests, eliminating any students with missing data. This results in only 399 students with complete data and the corresponding estimates are 0.68 (0.013) and −0.008 (0.015). In this case, even though the individual level 1

Table 6.4 Science attainment estimates for Hungary IEA study

(a) Fixed

	Estimate (s.e.)
Earth Science core	0.838 (0.0076)
Biology core	0.711 (0.0100)
Biology R3	0.684 (0.0109)
Biology R4	0.591 (0.0167)
Physics core	0.752 (0.0128)
Physics R2	0.664 (0.0128)
Earth Science core (girls–boys)	−0.0030 (0.0059)
Biology core (girls–boys)	−0.0151 (0.0066)
Biology R3 (girls–boys)	0.0040 (0.0125)
Biology R4 (girls–boys)	−0.0492 (0.0137)
Physics core (girls–boys)	−0.0696 (0.0073)
Physics R2 (girls–boys)	−0.0696 (0.0116)

(b) Random: variances on diagonal; correlations off diagonal

	E.Sc. core	Biol. core	Biol. R3	Biol. R4	Phys. core	Phys. R2
Level 2 (School)						
E.Sc. core	0.0041					
Biol. core	0.68	0.0076				
Biol. R3	0.51	0.68	0.0037			
Biol. R4	0.46	0.68	0.45	0.0183		
Phys. core	0.57	0.90	0.76	0.63	0.0104	
Phys. R2	0.54	0.78	0.57	0.65	0.78	0.0095
Level 1 (Student)						
E.Sc. core	0.0206					
Biol. core	0.27	0.0261				
Biol. R3	0.12	0.13	0.0478			
Biol. R4	0.14	0.27	0.20	0.0585		
Phys. core	0.26	0.42	0.11	0.27	0.0314	
Phys. R2	0.22	0.33	0.14	0.37	0.41	0.0449

correlations are relatively small, the gain in efficiency from carrying out the full multivariate analysis is substantial, especially for inferences about the gender difference which in the second analysis is less than its standard error.

Another way to combine the subtests would be to form, for each student, a score based upon the items which the student responded to. Thus, for Biology the 399 students taking the core and both rotated forms would have a score out of 18 items; and there would be 823 and 807 students respectively with scores out of 14 items with 410 students having only a score out of the core test. Since the scores are out of different totals, we would expect the between-student and between-school variances to differ and this is the case; the between-student variance for the 10 core test score is 0.00013 compared to that for the 18 item core and two rotated forms score of 0.00021. Thus, we would need to fit separate variance and covariance terms in general for each combination and in effect treat the four combinations as separate responses in order to obtain efficient estimates. Furthermore, we would also tend to obtain high correlations between these combination scores that could lead to numerical estimation problems, so that in general this procedure is not recommended.

6.5   Informative subject choice in examinations

In the previous example the choice of response was independent of the value of the response measurement since allocation of students was made at random. In some cases, however, this will not be true and we look at one such case where students taking an examination can make a choice of papers, and this choice is related to their underlying achievement.

Yang et al. (2001) studied a large group of students taking mathematics examinations at Advanced level in England in 1997, at age 18 years. There were 52 587 students in 2592 educational institutions. The data were analysed using a scoring system for achieved grades whereby the highest grade (A) is given a score of 10 and a failure is given a score of 0 with a mean overall of 6.4. Students take up to four papers chosen from 10 different types, with just under 87% choosing just one paper and just under 13% choosing two. It is clear from the data that students taking particular combinations of subjects tend to perform better than those taking other combinations. For example the average score on the main, basic mathematics paper is 5.54 for those who take this paper alone compared to 9.45 for those who take it in combination with a further mathematics paper. The analysis adjusts for prior achievement measured by the performance on the GCSE examination taken 2 years earlier and interest focuses on differences between institution types and on variability and correlations among papers at the school and student level.

In order to take account of informative choice the authors fit a term (dummy variable) for each combination of subjects taken and allow the coefficients of these to vary randomly at the school level and also for the student-level variance to depend on the combination chosen (see Chapter 3). They also allow interactions with other explanatory variables such as gender. They are able to conclude that the institutional level correlations among subjects, when different choice combinations are modelled, are typically only moderate and that institutions are more homogeneous with respect to results from some combinations as opposed to others. Models of this kind may be useful in other situations, for example where question choice within examinations is permitted.

6.6 Principal components analysis

We have already seen in section 6.1 that the covariance matrix for a multivariate response vector where there are missing data can be efficiently estimated by arranging for the multivariate structure to constitute a 'dummy' level 1. When the variables have a multivariate distribution the resulting estimates are maximum likelihood or restricted maximum likelihood.

The aim of principal components analysis is to find a linear function of a set of variates which has the maximum variance, subject to a suitable constraint. In the single level case we require to maximize the variance of $\mathbf{w}^T \mathbf{y}$ where $\mathbf{w}$ is the vector of weights defining the linear function of the variates $\mathbf{y}$, and Ω is the covariance matrix of $\mathbf{y}$, namely

$$\Lambda = \mathbf{w}^T \Omega \mathbf{w}, \quad \mathbf{w}^T \mathbf{w} = 1$$

The solution is given by the eigenvector associated with the largest eigenvalue of Ω, that is the solution of

$$|\Omega - \lambda I| = 0 \tag{6.2}$$

We define a second function by the set of weights that maximizes the variance subject to the function being uncorrelated with the first function. The solution is given by the eigenvector associated with the second largest eigenvalue, and subsequent functions can be defined similarly (Lawley and Maxwell, 1971). The variates are usually standardized to have equal variances.

We note that the covariance (or correlation) matrix Ω can be a residual matrix, after regressing on explanatory variables. Thus, if we wished to form a principal component for the four science subjects of the previous section, we may wish to use the residual covariance matrix, after adjusting for gender differences. We now, however, have a choice of two covariance matrices, the between-student and the between-school one. If we choose the between-student matrix, then we would interpret the principal component as that which had been adjusted for school differences. In forming the derived summary variable(s) we would not use the actual observed variates but the level 1 estimates of them, that is the level 1 residuals, the $\hat{u}_{01j}, \hat{u}_{02j}$ of (6.1).

We could also choose to summarize the level 3 covariance matrix, and in this case we would use the school-level residuals as the variates in the linear function. If the principal component analysis has been carried out on the residuals from a multivariate multilevel analysis then we may wish to regard the school-level principal component as a convenient summary measure of school differences.

Table 6.5 shows the student-level and school-level principal component weights for the Science data. Since the measures are designed to be on the same scale we work directly with the covariance matrices.

As might be expected, the components both have positive weights. At the school level, the percentage variation accounted for by the first component is high suggesting that school Science performance may usefully be summarized by this weighted function of the individual school-level subject residuals. Also, the two sets of weights are fairly similar. This suggests that if we wished to summarize the individual subject scores into a single index, we could do this using the student-level weights, or even the weights obtained using the total covariance matrix. In Chapter 7 we shall further explore the structure of these data using a factor analysis model.

Table 6.5   Principal component weights for science test scores and percentage variation accounted for

Subject	Between-student	Between-school
Earth Science core	0.17	0.21
Biology core	0.29	0.40
Biology R3	0.31	0.21
Biology R4	0.63	0.59
Physics core	0.35	0.46
Physics R2	0.52	0.43
% variation	41%	72%

6.7   Multiple discriminant analysis

Given a set of variates we can seek a linear function of them that best discriminates among groups and this leads to the following definition. If $\bar{\mathbf{y}}$ is the vector of group means then we require a set of weights $\mathbf{w}$ such that $\mathbf{w}^T\bar{\mathbf{y}}$ has maximum variance, subject to the within-group variance of $\mathbf{w}^T\mathbf{y}$ being constrained, for example equal to 1.0. The solution is the vector associated with the largest root of

$$|\Omega_B - \lambda\Omega_W| = 0$$

for the between-group (Ω_B) and within-group (Ω_W) covariance matrices. For just two groups this gives the usual 'Fisher' discriminant function. As in principal components analysis we can find further vectors that discriminate best, subject to being uncorrelated with all the previous vectors. The function of the variates $\mathbf{w}^T\mathbf{y}$ can then be used, for example, to classify a new unit into the 'nearest' group.

In the 2-level case our groups are the level 2 units so that we require the covariance matrices from both levels. Using the Science data example the first vector is given by the weights $0.41, -0.07, 1.00, 0.26, 0.31, 0.13$ and explains about 48% of the variation. The next two vectors account for 19% and 13%. It is difficult to interpret these weights and the function would seem to have limited usefulness in this case for discriminating between schools.

7

Multilevel Factor Analysis and Structural Equation Models

7.1   A two-stage 2-level factor model

The theory and application of single-level structural equation models, including the special cases of observed variable path models and factor analysis models, is well known (Joreskog and Sorbom, 1979; McDonald, 1985). We shall look at multilevel generalizations of these models, starting with a factor analysis model which is then elaborated. Early work on estimation procedures based upon maximum likelihood are set out in Goldstein and McDonald (1987), McDonald and Goldstein (1988) with elaborations by Muthen (1989) and Longford and Muthen (1992). Raudenbush (1995) applied the EM algorithm to estimation for a 2-level structural equation model and Rowe and Hill (1997) show how existing multilevel software can be used to provide approximations to maximum likelihood estimates in general multilevel structural equation models. We shall describe general approaches based both upon maximum likelihood and MCMC methods.

Consider first a basic 2-level factor model where we have, say, a set of measurements on each student within a sample of schools together with a set of measurements at the school level which may be aggregated student-level measurements. The response measurements of interest whose structure we wish to explore are assumed to be Normally distributed random variables. A further set of covariates, for example gender or social class, are explanatory variables which we may wish to condition on. For the p level 1 responses we first write a multivariate model with p responses, where in general some may be randomly missing.

$$y_{hij} = (X\beta)_{hij} + \sum_{h=1}^{p} e_{hij}z_{hij} + \sum_{h=1}^{p} u_{hj}z_{hij} \quad h = 1, \ldots, R \qquad (7.1)$$

where h indexes the response. This is fitted as a 3-level model as described in Chapter 6 with dummy variables z_{hij} for each response, that have random coefficients at level 2 and level 3. Note that at level 3 (between schools) some of the responses may not vary. Note also that in general some of the coefficients of the covariates may vary at level 3 and these would be incorporated as further level 3 random variables along with those above. Reverting to the original 2-level model we now have a set of level 1 random variables e_{hij} and a set of level 2 random variables u_{hj}. A general factor structure for the level 1 variables may involve factors defined at both level 1 and level 2, where we can write

$$e_{hij} = \sum_g \lambda_{1gh} f_{gij}^{(1)} + w_{hij}^{(1)}$$

$$u_{hj} = \sum_g \lambda_{2gh} f_{gj}^{(2)} + w_{hj}^{(2)}$$

(7.2)

for the factor structures at each level, where λ, f, w respectively refer to the loadings, factors and residuals or 'uniquenesses', and the superscripts indicate the levels of the data structure. We may wish to identify some of these factors as the 'same' factors at each level, for example by constraining certain loadings to be zero. In general of course, we may have different random variables at level 1 and level 2, since, for example, some of the variables which vary between students may not vary across schools and vice versa. Thus we may have an attitude score with no between-school variation and any aggregate-level variables by definition will not vary between pupils. The latter, nevertheless, may enter the model with the level 1 random variables as responses, by being part of the level 2 factor structure and contributing to the prediction of the u_{hj} in the above equation. Thus, we can in principle consider any level 2 random variables including random coefficients of covariates when modelling the factor structure at this level.

A straightforward and consistent procedure for estimating the parameters of this factor model is to do it in two stages. The first stage involves the estimation of the separate level 1 and level 2 residual covariance matrices as described above using the procedures given in Chapter 6. The second stage involves the factor analysis of these separate matrices using any standard procedure, as described for example in Joreskog and Sorbom (1979) or McDonald (1985). This also automatically deals with any missing responses at either level. McDonald (1993) gives details for maximum likelihood estimators in this case.

The two-stage procedure should be reasonably efficient except where the data are unbalanced, with highly variable numbers of level 1 units within level 2 units. It has the advantage that it can be used for quite general structures. Thus it extends straightforwardly to any number of hierarchical levels. Furthermore, we can also fit models where there are random cross-classifications using the procedures described in Chapter 11. Thus, if students are classified by the primary and the secondary school they attended we can estimate the covariance matrices for level 1 and for both classifications at level 2 and then carry out three separate factor analyses of these matrices. Rowe and Hill (1997) describe its application to some educational data.

This procedure also allows us to fit general unconditional path models, with or without latent variables, since the covariance matrices at each level are sufficient for these models. A simple example of such a model without latent variables is as follows

$$y_{ij}^{(1)} = \alpha_1 + \beta_1 x_{ij}^{(1)} + u_j^{(1)} + e_{ij}^{(1)}$$

$$y_{ij}^{(2)} = \alpha_2 + \beta_2 y_{ij}^{(1)} + u_j^{(2)} + e_{ij}^{(2)}$$

(7.3)

where the $y_{ij}^{(1)}$ is regarded as a random variable in both equations. The traditional path model treats $y_{ij}^{(1)}$ in the second of these equations conditionally, so that it can be treated straightforwardly as a bivariate 2-level model. A choice between these two models will depend on substantive considerations, especially where there is a temporal ordering of variables when the conditional model would seem to be more appropriate in general. McDonald (1985) gives an account of estimation for unconditional path models. We provide a more general specification of such models below.

7.2 A general multilevel factor model

We can write a general model with Normal responses as follows

$$y_{rij} = \beta_r + \sum_k \alpha_k x_{kij} + \sum_{f=1}^{F} \lambda_{fr}^{(2)} v_{fj}^{(2)} + \sum_{g=1}^{G} \lambda_{gr}^{(1)} v_{gij}^{(1)} + u_{rj} + e_{rij}$$

$$u_{rj} \sim N(0, \sigma_{ur}^2), \quad e_{rij} \sim N(0, \sigma_{er}^2), \quad v_{fj}^{(2)} \sim MVN_F(0, \Omega_2), \quad v_{gij}^{(1)} \sim MVN_G(0, \Omega_1)$$

$$(7.4)$$

$$r = 1, \ldots, R, \quad i = 1, \ldots, n_j, \quad j = 1, \ldots, J, \quad \sum_{j=1}^{J} n_j = N$$

Here we have R responses for N individuals split between J level 2 units. We have F sets of factors, $v_{fj}^{(2)}$ defined at level 2 and G sets of factors, $v_{gij}^{(1)}$ defined at level 1. In the fixed part of the model we fit separate intercept terms β_r for each response and allow covariates x_{kij}. The loadings, $\lambda^{(1)}, \lambda^{(2)}$, are specific to each level. It is possible to further elaborate this model by considering a factor structure for any random coefficients, α_{kj}, but we will not pursue this. The residuals, or 'uniquenesses' at levels 1 and 2, e_{rij} and u_{rj} are assumed to be mutually independent, that is with diagonal covariance matrices.

Rabe-Hesketh et al. (2001) discuss this model and extensions to structural equation models (see below) as well as models where the responses are discrete or mixtures of discrete and continuous responses with various link functions. They provide maximum likelihood procedures using quadrature estimation, which have been implemented for the STATA software package (see Chapter 15).

Although (7.4) allows a very flexible set of factor models it should be noted that in order for such models to be identifiable suitable constraints must be put on the parameters. These will consist of fixing the values of some of the elements of the factor variance matrices, Ω_1 and Ω_2, and/or some of the factor loadings, $\lambda_{fr}^{(2)}$ and $\lambda_{gr}^{(1)}$. See Everitt (1984) for a further discussion of identifiability.

If we knew the values of the loadings λ then we could fit (7.4) directly as a 3-level model with the loading vectors as the explanatory variables for level 2 and level 3 random coefficients with variances constrained to be equal to 1. Conversely, if we knew the values of the random effects v we could estimate the loadings as fixed coefficients in a multivariate response model. These considerations suggest that an EM algorithm can be used in the estimation where the random effects are regarded as missing data (see Rubin and Thayer, 1982). It also suggests an MCMC approach which has the advantage of providing exact inferences for parameters and which can incorporate prior information about parameters.

7.3   MCMC estimation for the factor model

To illustrate the procedure we consider the following very simple 1-level factor model first:

$$y_{ri} = \lambda_r v_i + e_{ri}, \quad r = 1, \ldots, R, \quad i = 1, \ldots, N$$

$$v_i \sim N(0, 1), \quad e_{ri} \sim N(0, \sigma_{er}^2)$$

(7.5)

We will assume that the factor loadings have Normal prior distributions, $p(\lambda_r) \sim N(\lambda_r^*, \sigma_{\lambda r}^2)$ and that the level 1 variance parameters have independent inverse Gamma priors, $p(\sigma_{er}^2) \sim \Gamma^{-1}(a_{er}^*, b_{er}^*)$. The * superscript is used to denote the appropriate parameters of the prior distributions. This model can be updated using a very simple three-step Gibbs sampling algorithm as shown below (see Appendix 2.4).

Step 1

Update $\lambda_r (r = 1, \ldots, R)$ from the following distribution: $p(\lambda_r) \sim N(\hat{\lambda}_r, D_r)$ where

$$D_r = \left(\frac{\sum_i v_i^2}{\sigma_{er}^2} + \frac{1}{\sigma_{\lambda r}^2} \right)^{-1}$$

and

$$\hat{\lambda}_r = D_r \left(\frac{\sum_i v_i y_{ri}}{\sigma_{er}^2} + \frac{\lambda_r^*}{\sigma_{\lambda r}^2} \right)$$

Step 2

Update $v_i (i = 1, \ldots, N)$ from the following distribution: $p(v_i) \sim N(\hat{v}_i, D_i)$ where

$$D_i = \left(\frac{\sum_r \lambda_r^2}{\sigma_{er}^2} + 1 \right)^{-1}$$

and

$$\hat{v}_i = D_i \left(\frac{\sum_r \lambda_r y_{ri}}{\sigma_{er}^2} \right)$$

Step 3

Update σ_{er}^2 from the following distribution: $p(\sigma_{er}^2) \sim \Gamma^{-1}(\hat{a}_{er}, \hat{b}_{er})$ where

$$\hat{a}_{er} = \frac{N}{2} + a_{er}^* \quad \text{and} \quad \hat{b}_{er} = \frac{1}{2} \sum_i e_{ri}^2 + b_{er}^*.$$

Goldstein and Browne (2002a) show how the MCMC algorithm can be extended for general multilevel models which allow both orthogonal and correlated factor structures using Gibbs and MH sampling. They also propose a goodness of fit statistic for model comparison. Ansari and Jedidi (2002) also give a general exposition of this model, provide a model comparison procedure, and show how a failure to model the multilevel structure can lead to misleading inferences.

Table 7.1   Science attainment 2-level factor model: MCMC estimates

Parameter	A Estimate (s.e.)	B Estimate (s.e.)
Level 1 factor 1 loadings		
Earth Sc. core	0.06 (0.004)	0.11 (0.02)
Biol. core	0.11 (0.004)	0*
Biol. R3	0.05 (0.008)	0*
Biol. R4	0.11 (0.009)	0*
Phys. core	0*	0*
Phys. R2	0*	0*
Level 1 factor 2 loadings		
Earth Sc. core	0*	0*
Biol. core	0*	0.10 (0.005)
Biol. R3	0*	0.05 (0.008)
Biol. R4	0*	0.10 (0.009)
Phys. core	0.12 (0.005)	0*
Phys. R2	0.12 (0.007)	0*
Level 1 factor 3 loadings		
Earth Sc. core	—	0*
Biol. core	—	0*
Biol. R3	—	0*
Biol. R4	—	0*
Phys. core	—	0.12 (0.005)
Phys. R2	—	0.12 (0.007)
Level 2 factor 1 loadings		
Earth Sc. core	0.04 (0.007)	0.04 (0.007)
Biol. core	0.09 (0.008)	0.09 (0.008)
Biol. R3	0.05 (0.009)	0.05 (0.010)
Biol. R4	0.10 (0.016)	0.10 (0.016)
Phys. core	0.10 (0.010)	0.10 (0.010)
Phys. R2	0.09 (0.011)	0.09 (0.011)
Level 1 residual variances		
Earth Sc. core	0.017 (0.001)	0.008 (0.004)
Biol. core	0.015 (0.001)	0.015 (0.001)
Biol. R3	0.046 (0.002)	0.046 (0.002)
Biol. R4	0.048 (0.002)	0.048 (0.002)
Phys. core	0.016 (0.001)	0.016 (0.001)
Phys. R2	0.029 (0.002)	0.030 (0.002)
Level 2 residual variances		
Earth Sc. core	0.002 (0.0005)	0.002 (0.0005)
Biol. core	0.0008 (0.0003)	0.0008 (0.0003)
Biol. R3	0.002 (0.0000)	0.002 (0.0008)
Biol. R4	0.010 (0.002)	0.010 (0.002)
Phys. core	0.002 (0.0005)	0.002 (0.0005)
Phys. R2	0.003 (0.0009)	0.003 (0.0009)
Level 1 correlation factors 1 & 2	0.90 (0.03)	0.55 (0.10)
Level 1 correlation factors 1 & 3	—	0.49 (0.09)
Level 1 correlation factors 2 & 3	—	0.92 (0.04)

*Indicates constrained parameter. A chain of length 20 000 with a burn in of 2000 was used. Level 1 is student, level 2 is school.

7.3.1   A two-level factor example

We use the Science attainment test score data described in Chapter 6 to fit two 2-level factor models to these data. The results are given in Table 7.1. We omit the fixed effects estimates since they are very close to those in Table 6.4. Model A has two factors at level 1 and a single factor at level 2. For illustration we have constrained all the variances to be 1.0 and allowed the covariance (correlation) between the level 1 factors to be estimated. Inspection of the correlation structure suggests a model where the first factor at level 1 estimates the loadings for Earth Science and Biology, constraining those for Physics to be zero (the physics responses have the highest correlation), and for the second factor at level 1 to allow only the loadings for Physics to be unconstrained. The high correlation of 0.90 between the factors suggests that perhaps a single factor will be an adequate summary. Although we do not present results, we have also studied a similar structure for two factors at the school level where the correlation is estimated to be 0.97, strongly suggesting a single factor at that level.

For model B we have separated the three topics of Earth Science, Biology and Physics to separately have non-zero loadings on three corresponding factors at the student level. This time the high inter-correlation is that between the Biology and Physics booklets with only moderate (0.49, 0.55) correlations between Earth Science and Biology and Physics. In fact this model is not properly identified because of the single loading for Earth Sciences Core at level 1. The correlation estimates are strongly dependent on the chosen prior, in this case for the correlations the prior is uniform over $(-1, 1)$ and a different choice of prior, for example uniform over $(0, 1)$ will lead to different estimates. In the extreme case a point prior is equivalent to fixing the value of the correlation to a given value. In general we can use informative, other than point, priors for any of the parameters in the model and this provides an alternative method of handling identifiability to that of fixing parameters at particular values.

The results suggest that we need at least two factors to describe the student-level data and that the preliminary analysis using just one factor can be improved upon.

7.4   Structural equation models

A fairly general, single-level, structural equation model can be written in the following matrix form (see McDonald (1985) for some alternative representations)

$$A_1 v_1 = A_2 v_2 + W$$
$$Y_1 = \Lambda_1 v_1 + U_1$$
$$Y_2 = \Lambda_2 v_2 + U_2 \tag{7.6}$$

where Y_1, Y_2 are observed multivariate vectors of responses, A_1 is a known transformation matrix, often set to the identity matrix, A_2 is a coefficient matrix which specifies a multivariate linear model between the set of transformed factors, v_1, and v_2, Λ_1, Λ_2 are loadings, U_1, U_2 are uniquenesses, W is a random residual vector and W, U_1, U_2 are mutually independent with zero means. The extension of this model to the multilevel case follows that of the factor model and we shall restrict ourselves to sketching how the MCMC algorithm can be applied to (7.6). Note that we can write

A_2 as the vector A_2^* by stacking the rows of A_2. For example if:

$$A_2 = \begin{pmatrix} a_0 & a_1 \\ a_2 & a_3 \end{pmatrix}, \quad \text{then } A_2^* = \begin{pmatrix} a_0 \\ a_1 \\ a_2 \\ a_3 \end{pmatrix}$$

The distributional form of the model can be written as

$$A_1 v_1 \sim MVN(A_2 v_2, \Sigma_3)$$
$$v_1 \sim MVN(0, \Sigma_{v_1}), \quad v_2 \sim MVN(0, \Sigma_{v_2})$$
$$Y_1 \sim MVN(\Lambda_1 v_1, \Sigma_1), \quad Y_2 \sim MVN(\Lambda_2 v_2, \Sigma_2)$$

with priors

$$A_2^* \sim MVN(\hat{A}_2^*, \Sigma_{A_2^*}), \quad \Lambda_1 \sim MVN(\hat{\Lambda}_1, \Sigma_{\Lambda_1}), \quad \Lambda_2 \sim MVN(\hat{\Lambda}_2, \Sigma_{\Lambda_2})$$

and $\Sigma_1, \Sigma_2, \Sigma_3$ having inverse Wishart priors.

The coefficient and loading matrices have conditional Normal distributions as do the factor values. The covariance matrices and uniqueness variance matrices involve steps similar to those given in the earlier algorithm. The extension to two levels and more follows readily.

The model can be generalized further by considering m sets of response variables, $Y_1, Y_2, \ldots, Y_m$ in (7.6), and several, linked, multiple group structural relationships with the k-th relationship having the general form

$$\sum_h V_h^{(k)} A_h^{(k)} = \sum_g V_g^{(k)} A_g^{(k)} + W^{(k)}$$

and the above procedure can be extended for this case. Note that the model for simultaneous factor analysis (or, more generally, structural equation model) in several populations is a special case of this model, with the addition of any required constraints on parameter values across populations.

We can also generalize (7.1) so that it includes fixed effects, responses at level 2 and covariates Z_1, Z_2 for the factors, which may be a subset of the fixed effects covariates X, namely

$$Y^{(1)} = X\beta + \Lambda_2^{(1)} v_2 Z_2^{(1)} + u^{(1)} + \Lambda_1^{(1)} v_1 Z_1 + e^{(1)}$$
$$Y^{(2)} = \Lambda_2^{(2)} v_2 Z_2^{(2)} + u^{(2)}$$

$$Y^{(1)} = \{y_{rij}\}, \quad Y^{(2)} = \{y_{rj}\}$$
$$r = 1, \ldots, R \quad i = 1, \ldots, i_j \quad j = 1, \ldots, J$$

(7.7)

The superscript refers to the level at which the measurement exists, so that, for example, y_{1ij}, y_{2j} refer respectively to the first measurement in the i-th level 1 unit in the j-th level 2 unit (say students and schools) and the second measurement taken at school level for the j-th school.

7.5 Discrete response multilevel structural equation models

Suppose we have a set of binary responses for a set of variables measured on individuals. We can write a model analogous to (7.5) as

$$f(\pi_{ri}) = \lambda_r v_i + e_{ri}, \quad r = 1, \dots, R, \quad i = 1, \dots, N$$

$$v_i \sim N(0, 1), \quad e_{ri} \sim N(0, \sigma_{er}^2) \tag{7.8}$$

$$y_{ri} \stackrel{iid}{\sim} \text{Bin}(1, \pi_{ri})$$

where for convenience we choose a probit link function so that the unique variances, σ_{er}^2, are fixed at 1. As before we can use a Gibbs sampler (see Appendix 4.3) where the lowest level variables are also thought of as latent, and generalizations to several levels and to structural equations follow similar lines to those described above. Model (7.8) is often referred to as an item response model (IRM) (see Goldstein and Wood, 1989) and (7.8) and its extensions allow a very flexible class of models to be fitted. Thus, for example, for an ordered response model, sometimes termed a partial credit model, we can also use the probit link and can therefore fit mixtures of binary, ordered and continuous responses in the same model. Goldstein and Browne (2002b) fit a 2-level binary response factor model with a probit link to a set of 15 Mathematics items from a test given to French and British school students. They show how country differences can be modelled and how multiple factors, for each country, at both levels can be fitted.

8

Nonlinear Multilevel Models

8.1 Introduction

The models of Chapters 2 and 3 are linear in the sense that the response is a linear function of the parameters in the fixed part and the elements of V are linear functions of the parameters in the random part. In many applications, however, it is appropriate to consider models where the fixed or random parts of the model, or both, contain nonlinear functions. For example, in the study of growth, Jenss and Bayley (1937) proposed the following function to describe the growth in height of young children

$$y_{ij} = \alpha_0 + \alpha_1 t_{ij} + u_{\alpha 0 j} + u_{\alpha 1 j} t_{ij} + e_{\alpha ij} - \exp(\beta_0 + \beta_1 t_{ij} + u_{\beta 0 j} + u_{\beta 1 j} t_{ij} + e_{\beta ij}) \quad (8.1)$$

where t_{ij} is the age of the j-th child at the i-th measurement occasion. Models for discrete data, such as counts or proportions, are a special case of nonlinear models often termed 'generalized linear models' and we discussed these in Chapter 4. For example, a 2-level log linear model can be written

$$E(m_{ij}) = \pi_{ij}, \quad \pi_{ij} = \exp(X_{ij}\beta_j) \quad (8.2)$$

where m_{ij} is assumed typically to have a Poisson distribution, in this case across level 1 units. In this chapter we consider nonlinear models in general.

8.2 Nonlinear functions of linear components

The following results are an extension of those presented by Goldstein (1991) and Appendix 8.1 gives details. We shall give procedures for deriving likelihood-based estimates and note that where the random variables are not part of the nonlinear function, the procedure gives maximum likelihood estimates; otherwise these are quasilikelihood estimates similar to those considered in Chapter 4.

MCMC methods can also be used to fit these models. As with generalized linear models (see Appendix 4.3) MH sampling can be used since not all the conditional distributions can be specified analytically. Otherwise the steps are as in Appendix 2.4. The likelihood-based estimates described here can be used to provide starting values for an MCMC estimation.

Restricting attention to a 2-level structure we can write a fairly general model as follows

$$y_{ij} = X_{1ij}\beta_1 + Z_{1ij}^{(2)}u_{1j} + Z_{1ij}^{(1)}e_{1ij} + f(X_{2ij}\beta_2 + Z_{2ij}^{(2)}u_{2j} + Z_{2ij}^{(1)}e_{2ij}) + \dots \quad (8.3)$$

where the function f is nonlinear and where the $+\ldots$ indicates that additional nonlinear functions can be included, involving further fixed part explanatory variables X or random part explanatory variables at levels 1 and 2, respectively $Z^{(1)}$, $Z^{(2)}$. The model is first linearized by a suitable Taylor series expansion and this leads to consideration of a linear model where the explanatory variables in f are transformed using first and second derivatives of the nonlinear function. Note that the linear component of (8.3) is treated in the standard way, and that the random variables at a given level in the linear and nonlinear components may be correlated.

Consider the nonlinear function f. Appendix 8.1 gives details to show that we can write this as the sum of a fixed part component and a random part using a Taylor series approximation. Concentrating on the fixed part this gives

$$f_{ij}(H_{t+1}) = f_{ij}(H_t) + X_{ij}(\beta_{1,t+1} - \beta_{1,t})f'_{ij}(H_t) \qquad (8.4)$$

where $\beta_{1,t+1}$, $\beta_{1,t}$ are the current and previous iteration values of the fixed part coefficients and H_t represents the fixed part predictor in the nonlinear component of (8.3). We can choose H_t to be either the current value of the fixed part predictor, that is $X_{2ij}\beta_2$, or we can add the current estimated residuals to obtain an improved approximation to the nonlinear component for each unit. As with generalized linear models the former is referred to as a 'marginal' (quasilikelihood) model and the latter as a 'penalized' or 'predictive' (quasilikelihood) model. We note that this procedure reduces to the standard Newton–Raphson scoring algorithm for a single level model (McCullagh and Nelder, 1989).

In practice general models such as (8.1) may pose considerable estimation problems. We notice that the same explanatory variables can occur in the linear and nonlinear components and this can lead to instability and failure to converge.

8.3 Estimating population means

Consider the expected value of the response for a given set of covariate values. Because of the nonlinearity this is not in general equal to the predicted value when the random variables in the nonlinear function are zero. For example, if we write the variance components version of (8.2)

$$\pi_{ij} = \exp(\beta_0 + \beta_1 x_{ij} + u_j)$$

and assuming $u_j \sim N(0, \sigma_u^2)$ we obtain the marginal expectation

$$E(\pi_{ij}|x_{ij}) = \exp(\beta_0 + \beta_1 x_{ij}) \int_{-\infty}^{\infty} e^{u_j}\phi(u_j)du_j = \exp(\beta_0 + \beta_1 x_{ij} + \sigma_u^2/2) \qquad (8.5)$$

where ϕ is the density function of the Normal distribution. The population predicted values, conditional on covariates, can be obtained if required, as above, by taking expectations over the population. An approximation to this can be obtained from the second order terms in (8.1.4) in Appendix 8.1 with higher order terms introduced if necessary to obtain a better approximation. Alternatively we may generate a large number of simulated sets of values for the random variables and for each set evaluate the response function to obtain an estimate of the full population distribution. See section 4.8 for more details of this procedure.

8.4   Nonlinear functions for variances and covariances

We saw in Chapter 3 how we could model complex functions of the level 1 variance. As with the linear component of the model, there are cases where we may wish to model variances or covariances as nonlinear functions. In principle we can do this at any level but we restrict our attention to level 1 and to the variance only. In Chapter 5 we give an example where the covariances are modelled in this way.

Suppose that the level 1 variance decreases with increasing values of an explanatory variable such that it approaches a fixed value asymptotically. We could then model this for a 2-level model, say, as follows

$$\text{var}(e_{ij}) = \exp(\beta_0^* + \beta_1^* x_{ij})$$

where β_0^*, β_1^* are parameters to be estimated. Such a model also guarantees that the level 1 variance is positive which is not the case with linear models, such as those based on polynomials. The estimation procedure is analogous to that described above and details are given in Appendix 8.1.

8.5   Examples of nonlinear growth and nonlinear level 1 variance

We give first an example of a model with a nonlinear function for the linear component and we then consider the case of a nonlinear level 1 variance function.

We use an example from child growth, consisting of 577 repeated measurements of height on 197 French Canadian boys aged from 5 to 10 years (Demirjian et al., 1982) with between 3 and 7 measurements each. This is a 2-level structure with measurement occasions nested within children. We fit the following version of the Jenss–Bayley curve to illustrate the procedure

$$y_{ij} = \alpha_0 + \exp(\beta_0 + \beta_1 t_{ij} + \beta_2 t_{ij}^2 + \beta_3 t_{ij}^3 + u_{\beta 0 j} + u_{\beta 1 j} t_{ij}) + e_{\alpha ij} \qquad (8.6)$$

so that the fixed part is an intercept plus a nonlinear component and the random part variance at level 2 is part of the nonlinear component. The results are given in Table 8.1 using the first order approximation with prediction based upon the fixed part only.

The level 1 variance is small and of the order of the measurement error of height measurements. The starting values for this model need to be chosen with care, and in the present case the model was run to convergence without the linear intercept α_0 which was then added with a starting value of 100. In another example, Bock (1992) uses an EM algorithm to fit a nonlinear 2-level model to growth data from age 2 years to adulthood using a mixture of three logistic curves.

The second example uses the JSP dataset where we studied the level 1 variance in Chapter 3. We will fit model B of Table 3.1 with a nonlinear function of the level 1 variance instead of the level 1 variance as a quadratic function of the 8-year-score. This level 1 variance for the ij-th level 1 unit is $\exp(\beta_0^* + \beta_1^* x_{1ij})$ and Table 8.2 shows

Table 8.1   Nonlinear model estimates with first order fixed part prediction. Age is measured about 8.0 years

(a) Fixed coefficient

	Estimate (s.e.)
Intercept (linear)	90.3
Intercept (nonlinear)	3.58
Age	0.15 (0.10)
Age squared	−0.016 (0.02)
Age cubed	0.002 (0.004)

(b) Nonlinear model level 2 covariance matrix (s.e.)

	Intercept	Age
Intercept	0.025 (0.003)	
Age	−0.0027 (0.0003)	0.00036 (0.00005)

Level 1 variance $= 0.25$.

Table 8.2   Nonlinear level 1 variance for JSP data

Parameter	Estimate (s.e.)
Fixed	
Constant	31.7
8-year score	0.58 (0.03)
Gender (boys–girls)	−0.34 (0.27)
Social class (non-manual–manual)	0.76 (0.30)
School mean 8-year score	0.01 (0.11)
8-year score × school mean 8-year score	0.02 (0.01)
Random	
Level 2	
σ_{u0}^2	2.87 (0.88)
σ_{u01}	−0.17 (0.07)
σ_{u1}^2	0.012 (0.007)
Level 1	
β_0^*	2.74 (0.06)
β_1^*	−0.10 (0.01)

the model estimates. In fact the estimates are almost identical to those of model B of Table 3.1 as is the likelihood value.

Figure 8.1 shows the predicted level 1 variance for this model and model B of Table 3.1.

In these data the nonlinear function gives very similar results to the quadratic one. It is clear, however, that where the variance asymptotically approaches a constant value, for extreme values of an explanatory variable, a linear or even quadratic approximation may be expected to fail. In the present case a linear function does predict a negative level 1 variance within the range of the data. An example where a nonlinear function is necessary is in growth data, described in Chapter 5, where the level 1 (within-individual) variation will decrease towards a constant value at the approach to adulthood.

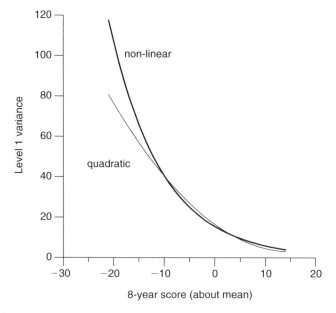

Figure 8.1 Predicted level 1 variance as a function of 8-year Maths score.

8.6 Multivariate nonlinear models

We can use the procedures of this chapter to fit multivariate models by using level 1 to define the multivariate structure and using the linearization procedures described in this chapter for higher levels. In general, the response variables will have different nonlinear link functions, and some of them may be linear. In Chapter 4 we discussed such a model where one response, say a mathematics test score, is a linear function of explanatory variables and a second response, say whether or not the student has a 'positive' attitude towards mathematics, is binary. For each level we will have variances for each response and covariances among the coefficients random at that level, where these are specified for the transformed model.

We may also have multivariate models where the level 1 variances are different nonlinear or linear functions of explanatory variables with covariances between the coefficients in the different nonlinear or linear functions.

Appendix 8.1

Nonlinear model estimation

The following exposition is similar to that in Appendix 4.1 for discrete response (non-linear) models, but is more general since the level 1 random effects are also part of the nonlinear link function. We consider a single nonlinear term of the form

$$y_{ij} = f(X_{ij}\beta_2 + Z_{2ij}u_{2j} + Z_{1ij}e_{1ij}) \tag{8.1.1}$$

The addition of linear terms to this model is discussed in Chapter 8.

At the $(t + 1)$-th iteration we expand (8.1.1) for both fixed and random parts as follows:

$$
\begin{aligned}
&f_{ij}(H_t) + X_{ij}(\beta_{t+1} - \beta_t)f'_{ij}(H_t) \\
&+ (Z_{2ij}u_{2j} + Z_{1ij}e_{1ij})f'_{ij}(H_t) + (Z_{2ij}u_{2j} + Z_{1ij}e_{1ij})^2 f'_{ij}(H_t)/2
\end{aligned} \tag{8.1.2}
$$

in terms of parameter values estimated at the t-th iteration. The first line of (8.1.2) updates the fixed part of the model and in the special case of a single-level quasi-likelihood model provides the updating function. The quantity $f_{ij}(H_t) - X_{ij}\beta_t f'_{ij}(H_t)$ is treated as an offset to be subtracted from the response variable. The first term in the second line defines a linear random component based on the explanatory variables transformed by multiplying by the first differential. We need to specify H_t and consider the distribution of the second term in the second line of (8.1.2).

If we choose $H_t = X_{ij}\beta_t$, this is equivalent to carrying out the Taylor expansion around the fixed part predicted value. If we choose $H_t = X_{ij}\beta_t + Z_{2ij}\hat{u}_{2j} + Z_{1ij}\hat{e}_{1ij}$, this expands around the current predicted value for the ij-th unit and we replace the second line of (8.1.2) by

$$
\begin{aligned}
&(Z_{2ij}(u_{2j} - \hat{u}_{2j}) + Z_{1ij}(e_{1ij} - \hat{e}_{1ij}))f'_{ij}(H_t) \\
&+ (Z_{2ij}(u_{2j} - \hat{u}_{2j}) + Z_{1ij}(e_{1ij} - \hat{e}_{1ij}))^2 f''_{ij}(H_t)/2
\end{aligned}
$$

We thus have the further offset from the linear term to be added to the response

$$(Z_{2ij}\hat{u}_{2j} + Z_{1ij}\hat{e}_{1ij})f'_{ij}(H_t)$$

From the second line of (8.1.2), where the Taylor expansion is about zero, we have

$$
\begin{aligned}
&E(Z_{2ij}u_{2j} + Z_{1ij}e_{1ij}) = 0, \quad E(Z_{2ij}u_{2j} + Z_{1ij}e_{1ij})^2 = \sigma_{zu}^2 + \sigma_{ze}^2 \\
&\sigma_{zu}^2 = Z_{2ij}\Omega_u Z_{2ij}^T, \quad \sigma_{ze}^2 = Z_{1ij}\Omega_e Z_{1ij}^T
\end{aligned} \tag{8.1.3}
$$

To incorporate the second order terms we treat $(\sigma_{zu}^2 + \sigma_{ze}^2)f''(H_t)/2$ as an additional offset in the fixed part and in the random part of the model we need to consider the variation of the second term in the second line of (8.1.2). If we assume Normality then all third moments, formed from the product of the two terms in the second line of (8.1.2), are zero and we have

$$\mathrm{var}(Z_{2ij}u_{2j} + Z_{1ij}e_{1ij})^2 = 2(\sigma_{zu}^4 + \sigma_{ze}^4) \tag{8.1.4}$$

so that we need to define the additional random variables

$$Z_u^* = \sigma_{zu}^2 f''(H_t)/\sqrt{2}, \quad Z_e^* = \sigma_{ze}^2 f''(H_t)/\sqrt{2}$$

which are uncorrelated and with variances constrained to be equal to 1.0. Equivalently we can form $Z_u^* Z_u^{*T}$, $Z_e^* Z_e^{*T}$ as offsets for the response vector $vec(\tilde{Y}\tilde{Y}^T)$ in the estimation of the random parameters. Having modified the response variable by removing the necessary offsets we are left in the fixed part with a modified response, say Y' with a modified explanatory variable matrix, say X'. We do likewise for the random part of the model and then carry out a standard iterative procedure, updating the differential functions at each iteration.

Where the Taylor expansion is taken about the current values of the residuals we require

$$\mathrm{E}[Z_{2ij}(u_{2j} - \hat{u}_{2j})]^2 + E[Z_{2ij}(e_{1ij} - \hat{e}_{1ij})]^2$$

which leads to the 'conditional' or 'comparative' variances described in Appendix 2.2, so that we substitute these variances, $\Omega_{\hat{u}}$ and $\Omega_{\hat{e}}$, for Ω_u and Ω_e, in the above expressions for the fixed and random offsets.

To estimate residuals we note that, having adjusted the response using the offsets, we have on the right-hand side of the model, for the Taylor expansion about zero, the fixed part together with the random terms.

$$(Z_{2ij}u_{2j} + Z_{1ij}e_{1ij})f'_{ij}(H_t) + [(Z_{2ij}u_{2j} + Z_{1ij}e_{1ij})^2 - (\sigma_{zu}^2 + \sigma_{ze}^2)]f''_{ij}(H_t)/2$$

Each residual and its square appear in this expression, and since third order moments are zero, we can apply the usual linear estimation for the residuals as described in Appendix 2.2. The weight matrix V is based upon both the linear and quadratic terms of the above expression. We carry out an analogous procedure for the case where the Taylor expansion is based upon the current residual estimates.

The above can be extended in a straightforward way to more than two levels and of course to multivariate models.

8.1.1 Modelling variances and covariances as nonlinear functions

In Chapter 2 we saw that the random parameters were estimated by regressing the observed cross-product matrix of residuals on a set explanatory variables which defined the appropriate variances and covariances at each level. Using the notation in Appendix 2.2 we have the following *linear* model for the random parameters β^*

$$Y^* = vec(\tilde{Y}\tilde{Y}^T) = X^*\beta^*, \quad E(Y^*) = vec(V) \tag{8.1.5}$$

We can now apply the same procedure for the specification and estimation of a nonlinear model as above. We illustrate this for the case where the level 1 variance is an exponential function of a covariate X_1^*, defined in terms of the Kronecker product as in Appendix 2.1, namely for the t-th element of $X^*\beta^*$ (which is on the diagonal of V); the level 1 variance contribution is

$$\sigma_{et}^2 = f(\beta^*) = \exp(\beta_0^* x_{0t}^* + \beta_1^* x_{1t}^*), \quad X_1^* = \{x_{1t}^*\}, \quad \beta^* = \begin{pmatrix} \beta_0^* \\ \beta_1^* \end{pmatrix} \quad (8.1.6)$$

As in the linear function case we form the first differential $f' = f$, multiply x_{0t}^*, x_{1t}^* by this and estimate the parameters of the resulting transformed linear model. This will involve introducing an offset for Y^* and constructing the following level 1 explanatory variables for the estimation of β^*, setting their covariance to zero

$$\{x_{0t}^* \exp(\beta_0^* x_{0t}^* + \beta_1^* x_{1t}^*)\}^{0.5}, \quad \{x_{1t}^* \exp(\beta_0^* x_{0t}^* + \beta_1^* x_{1t}^*)\}^{0.5}$$

Because we are estimating only nonlinear functions of linear components here and not adding approximations to a further random component, the estimates obtained are exact maximum likelihood or restricted maximum likelihood estimates.

In Chapter 5 we developed a special case of a nonlinear model for covariances and in Chapter 8 we give an example of model (8.1.6). We note that the parameters β_0^*, β_1^* are not necessarily positive when modelling (8.1.6) and although we would normally regard such level 1 parameters as variances, in this case as in section 3.1 they are simply parameters to be estimated. As with nonlinear modelling in general it is important to have reasonable starting values. These might be obtained by trial and error or by making preliminary estimates of variances for various values of the relevant explanatory variable and regressing their logarithms on the level 1 explanatory variables.

8.1.2 Likelihood values

The log-likelihood for the general multilevel model, apart from a constant and assuming multivariate Normality, can be written as

$$\log L = -tr(V^{-1}S) - \log|V|, \quad S = \tilde{Y}\tilde{Y}^T, \quad \tilde{Y} = Y - X\beta \quad (8.1.7)$$

An approximation to this for nonlinear models of a linear component is given by substituting the nonlinear function $f(X\beta)$ for $X\beta$ in (8.1.7) with the transformed random parts of the model incorporated into V in the usual way. If we use the predicted residuals to form H_t then we omit these from the likelihood calculation but add the offset term defined in (8.1.2) to $X\beta$. Likewise, in the second order model, we have to add the corresponding offsets to V. This procedure is equivalent to computing the ordinary likelihood using the modified response and explanatory variables Y', X' at convergence.

The estimates of $-2 \log L$ computed in this way can be used for approximate tests of hypotheses and for constructing confidence intervals, although the approximations may not be very good in some situations. When modelling variances and covariances as nonlinear functions the estimates obtained are in fact maximum likelihood ones and we obtain the exact value of $-2 \log L$.

9

Multilevel Modelling in Sample Surveys

9.1 Sample survey structures

Most practical sample surveys have a multistage sample design structure whereby at the first stage Primary Sampling Units (PSUs) are chosen randomly from a population of such units. Thus, in a household survey these might be large administrative areas. At the next stage either all the units or a sample of units is drawn from the chosen PSUs and this can lead to a third stage etc. until the final stage units, the households or persons in this case, are selected. This procedure is often referred to as 'cluster sampling', although strictly this term refers to samples where all the final stage units from each penultimate cluster are chosen. Typically, the selection probabilities at each stage are chosen so that the final sample is *self weighting*, that is, each final stage unit in the population has the same overall probability of being chosen. In a later section we shall look at how the selection probabilities at each stage can be used in model estimation. In this chapter we shall assume that there are no missing data; see Chapter 14 for procedures to handle such cases.

Sample surveys also often involve stratification; for example the population might be divided into rural and urban areas and sampling carried out separately for these 'strata', again often so that a self-weighting sample is produced. In some cases stratification may be used to ensure sufficient numbers in a particular category, such as an ethnic minority group. When we come to model a stratified sample we may wish to include in the model some dummy variable terms for the strata categories, and possibly for the categories formed by combinations of strata. Alternatively we may wish to incorporate stratum weights into the overall survey weights for each unit. We might wish to do this if we specify a 'marginal' model that describes the population as a whole rather than describing relationships within strata.

An important reason for stratification and clustering is to increase precision for a given cost for a given total sample size. Clustering, for example by area, generally reduces survey costs while increasing standard errors of estimates, while stratification tends to reduce standard errors (see for example Kish, 1965), and surveys are typically concerned to achieve a judicious balancing of these two procedures. From a modelling perspective, however, this is of secondary importance since we wish to uncover relationships that exist and estimate covariance structures. Where multistage surveys choose units for administrative and cost-effectiveness reasons, there will usually be little intrinsic interest in the between and within unit variation, but it will still usually be efficient to fit a multilevel model. The traditional approach in survey analysis to making inferences about the population, which ignores the existence of any multilevel structure, is to fit models appropriate for single level structures and then to adjust the standard errors to take account of the sample structure (see for example Kish and Frankel, 1974). Thus, for example, we saw in section 2.7 that using an OLS estimator when there were cluster differences tended to underestimate the fixed effect standard errors. It is also worth pointing out that, even where a survey involves no clustering or stratification, we may still wish to fit a multilevel model to explore the population data structure.

9.2   Population structures

9.2.1   Superpopulations

A major distinction is between inferences made for finite populations, and those for infinite populations where we make inferences on the basis of standard statistical model assumptions. Much of traditional sample survey theory is concerned with sampling from real finite populations where we wish to estimate characteristics of the population, such as means or proportions. Thus, in a household survey of a large city we may wish to estimate the actual proportion of one-parent households in that city or in each area of the city, for administrative reasons. However, for scientific purposes we will typically wish to consider the actual population sampled *as if* it was a realization of a conceptually infinite population extending through time, and also possibly through space. By taking this latter view we are able to make generalizations and predictions beyond the units that comprise the real population that has actually been sampled according to a specified sampling scheme.

If we adopt this conceptualization, often termed a 'superpopulation' or 'model-based' approach, then the use of multilevel modelling becomes a natural way to proceed. Nevertheless, we do need to consider whether the sampling selection process is relevant to the modelling. Generally, if the sampling process (clustering and stratification) is properly incorporated within the model, as described above, we need pay no further attention to the sample selection process, for instance the weights associated with each unit. Even where we do incorporate terms for stratification and clustering, however, there may still be an issue as to whether this has been done adequately, for example whether all possible interactions and random coefficients are included: we may still wish therefore to incorporate elements of the design into the analysis.

We shall assume for the present that any sample clustering, where there is variation between clusters, is modelled, although all aspects of the sample design may not be incorporated fully into the model. We look first at the case where the design is independent of the response variable values. We assume that the sampling scheme generates a set of sample selection probabilities for each unit at each level of the modelled hierarchy. For a 2-level model this will be the set $\{\pi_j, \pi_{i|j}\}$ and this then provides corresponding sampling weights $\{\pi_j^{-1}, \pi_{i|j}^{-1}\}$. We now carry out an analysis using these weights as described in section 3.4. Thus the model estimates take account of the sample design and such a weighted analysis ensures consistent estimates even where the model does not fully incorporate the sample design, whether this is by mistake or deliberately.

The second case is where the sample design depends on the values of the response, for example by stratifying on the basis of the mean value of the response variable for a geographical area, obtained for example from Census data. This is a case of 'informative sampling'. If the stratification is modelled, for example using dummy variables, we will have a model where one or more explanatory variables is correlated with one or more random effects. Another example is where the response is associated with the size of a sampling unit and hence with the selection probability. For such cases we will therefore need to use procedures such as those described in section 14.7. A detailed investigation of different weighting procedures here is given by Pfeffermann et al. (1997). They show that the procedure described in section 3.4 may not work well in this situation and propose a 'pseudo-likelihood' estimator instead, based around a modification of the IGLS algorithm.

9.2.2 Finite population inference

In the case of finite populations the traditional approach of 'design-based inference' aims to make inferences about the finite population quantities, for example means or proportions. These may be computed as simple (weighted) functions of the observations with standard errors constructed to reflect the sample design, and we may wish to provide these for the total population or for subpopulations or 'domains'; these domains may coincide with clusters or strata. So called 'model-assisted' inference extends this by using measured 'auxiliary' variables (X) in a regression model. Suppose that these auxiliary variables are available for every population unit and we have a 2-level model where the higher level units are clusters. For the sample we first define the 'multilevel synthetic estimate' using, for example, a simple variance components model such as

$$y_{ij}^{(s)} = (X^{(s)}\beta)_{ij} + u_{0j} + e_{ij} \qquad (9.1)$$

where the superscript (s) indicates that the observation is a member of the sample. Using the estimated regression coefficients β we can now form a prediction for every member of the population

$$\hat{y}_{ij} = (X\hat{\beta})_{ij} + \hat{u}_{0j} \qquad (9.2)$$

A synthetic estimate is then simply a suitable combination of these; for example the total for domain j is given by $\hat{y}_j = \sum_i \hat{y}_{ij}$. This estimator, however, is 'design-biased', that is it provides biased estimates with respect to repeated sampling from the same finite population. To correct for this bias a generalized regression (GREG) estimator can be used (Sarndal et al., 1992). In the present case we apply the GREG estimator to a multilevel model, hence known as the MGREG estimator (Lehtonen and Veijanen, 1999), to give

$$\hat{y}_j = \sum_{i \in D_j} \hat{y}_{ij} + \sum_{i \in D_j} w_{ij}(y_{ij}^{(s)} - \hat{y}_{ij}^{(s)}) \qquad (9.3)$$

where D_j refers to domain j and the w_{ij} are the sample weights, being the inverse of the selection probabilities within the domain, possibly adjusted for non-response. The second term on the right-hand side of (9.3) involves in this case the set of level 1 residuals, and essentially adjusts the bias in (9.2), at least approximately.

Clearly (9.1) and (9.3) can be extended to incorporate random coefficients and more complex multilevel structures. Such models can be expected to be more efficient than non-model-assisted estimators. This approach can incorporate nonlinear models, for example a logistic for binary responses; in this case the synthetic estimator consists of predicted probabilities. A further extension is to multivariate responses with appropriate modifications.

9.3   Small area estimation

A feature of many surveys is that they incorporate a large overall sample but for many domains or areas of interest the numbers are small, so that using just the sample members within such a domain will result in estimates with large standard errors. Thus a large household survey may select households in a large number of administrative districts, where the sample size within each district is then too small on its own to provide accurate estimates for each district. We shall consider such 'small area estimation' problems for a superpopulation model, and for finite populations we can modify the procedures as described above.

The simplest approach to this problem is to fit a multilevel model such as (9.1) and simply estimate the predicted 'synthetic' domain values derived from (9.2). In practice we will often have several surveys sharing some common domains and we look at how these can be used in combination to provide domain estimates that are as efficient as possible.

Suppose we have several data sets $T_1, \ldots, T_p$ sharing a common response and also sharing some of the same domains, the higher level area units, for example administrative regions. We shall consider the case of a single response to begin with and consider the extension to multiple responses later. Interest will typically lie in predicting domain means or totals. A special case is where one of the datasets has a special status, for example a population census or a set of administrative records. In such cases we may be able to treat the population estimates from this dataset as covariates in the model for the responses from the other surveys.

We shall use a variance components formulation, but the extension to random coefficients is straightforward, as is the extension to more complex models with further levels and cross-classifications for example.

For dataset T_h we have a 2-level linear variance components model

$$y_{ij}^{(h)} = (X^{(h)}\beta^{(h)})_{ij} + u_j^{(h)} + e_{ij}^{(h)} \tag{9.4}$$

We can allow the fixed and random part explanatory variables to be different in each dataset. Denote the set of domains for the overall model as S_1.

At the domain or area level (2) we have the set of (intercept) random effects

$$
\begin{aligned}
u_1^{(1)}, u_2^{(1)} \ldots \ldots u_1^{(2)}, \ldots \ldots, : \\
\text{cov}(u_j^{(h)}, u_{j'}^{(h')}) = 0, \quad j \neq j', \quad \text{cov}(u_j^{(h)}, u_j^{(h')}) = \sigma_{u(hh')}
\end{aligned}
\tag{9.5}
$$

Thus, by allowing the effects for higher level units to be correlated across surveys we are able to use all the available information efficiently to provide estimates for each domain. A number of special cases and extensions are of interest.

9.3.1 Information at domain level only

Suppose we have a number of domains (higher level units) where only domain level information is available. Such information might consist of administrative records. A variance components model for this set of domains can be written as

$$y._j = (X\beta)_j + u_j + e._j \qquad (9.6)$$

Denote the set of domains for this model by S_2. We have seen how such models can be fitted in Section 3.7. If we now form the union of this set of domains and the previous set, S_1, say $S_3 = \cup(S_1, S_2)$, this enhanced set can be analysed as a single model incorporating responses at different levels (section 3.6.1). A particular case of interest where the inclusion of such data can improve the estimates is for aggregate Census data, and we note again that the predictors in the component models can be different.

9.3.2 Longitudinal data

Repeated measurements over time can provide information which both increases efficiency and also allows the possibility of estimating trends so that estimates can be updated. Consider a single repeated survey where the same domains are measured across time with different level 1 units, say households, sampled at each occasion. A simple model would consist of a model, with a set of domains S_4, given by

$$y_{ijt} = (X\beta)_{ijt} + \alpha_0 + \alpha_1 t + u_{0j} + u_{1j}t + e_{ijt}$$
$$e_{ijt} \overset{iid}{\sim} N(0, \sigma_e^2), \quad \begin{pmatrix} u_{0j} \\ u_{1j} \end{pmatrix} \sim N(0, \Omega_u) \qquad (9.7)$$

which incorporates a time trend term in the fixed part and also allows the trend to vary across domains. In some cases, for example where population census data are available, it may be efficient to condition on the values of the response variable at a prior occasion. In (9.7) both the intercept and slope terms may covary with the domain level random effects for S_3, giving the combined model $S_5 = \cup(S_3, S_4)$ which can be fitted as a single model.

A particular problem with longitudinal data is mobility where individual people, or households, change domains during the course of a study. In principle such cases can be handled using multiple membership models (see Chapter 12). Where time is modelled in such studies, taking account of such mobility also implicitly allows migration patterns to influence the estimates. Where migration is high and 'informative' in the sense that those who move are a non-random subsample, taking account of such migration will be important in order to provide unbiased estimates.

9.3.3 Multivariate responses

Models with more than one response variable can be fitted as straightforward extensions to the above models. A particular advantage of such multivariate models is that we can obtain efficient estimates when one of the responses is missing, either completely at random or by design as in the rotation designs considered in Chapter 6. Thus, if information on one response is not included in some datasets by design we can still provide efficient estimates via the correlations with other responses. This will often be a more convenient and flexible method for exploiting these correlations than including the latter responses as covariates. Longford (1999) also addresses this issue and demonstrates the improved efficiency which can result.

10

Multilevel Event History Models

10.1 Introduction

This class of models has as the response variable the length of time between 'events'. An example is the beginning and end of a period of employment with the corresponding time being the duration of employment. The term 'survival model' is used when the end event is terminal such as a death, so that, unlike general event history models, repeated episodes are not possible. There is a considerable theoretical and applied literature, especially in the field of biostatistics, and a useful summary is given by Clayton (1988) and Singer and Willett (2002) provide illustrative applications. We consider first two basic approaches to the modelling of duration data. These are based upon 'proportional hazard' models and upon direct modelling of the log duration, often known as 'accelerated life models'. In both cases we can include explanatory variables. We then discuss discrete time models which are particularly suitable for fitting multilevel data structures.

The multilevel structure of such models arises in two general ways. The first is where we have repeated durations within individuals, analogous to our repeated measures models of Chapter 5. Thus, individuals may have repeated spells of various kinds of employment of which unemployment is one. In this case we have a 2-level model with individuals at level 2, often referred to as a renewal process. The second kind of model is where we have a single duration for each individual, but the individuals are grouped into level 2 units. In the case of employment duration the level 2 units could be firms or employers. If we had repeated measures on individuals within firms then this would give rise to a 3-level structure. We can further extend these models where there are cross-classifications (see Chapter 11) and multiple membership structures (see Chapter 12).

10.2 Censoring

A characteristic of duration data is that for some observations we may not know the exact duration but only that it occurred within a certain interval, which is known as interval censored data. Where the start of an interval occurred prior to the observation period it is known as left censored data, and where it occurred after a known period of observation, for example by an individual leaving a study, it is known as right censored data. For example, if we know at the time of a study only that someone entered her present employment before a certain date, then the information available is left censored and the duration is longer than a known value. In another case we may know that someone entered and then left employment between two measurement occasions, in which case we know only that the duration lies in a known interval. The models described in this chapter have procedures for dealing with interval and right censored data. In the case of the parametric models, where there are relatively large proportions of censored data, the assumed form of the distribution of duration lengths is important, whereas in the partially parametric models the distributional form is ignored. It is assumed that the censoring mechanism is non-informative, namely that the probability of censoring is independent of the actual duration length. Left censored data pose more of a problem and a standard procedure is to ignore such intervals where possible. Thus, for example, in repeated measures data for durations the first interval, if left censored, would be ignored.

In some cases, we may have data which are censored but where we have no duration information at all. For example, if we are studying the duration of first marriage and we end the study when individuals reach the age of 30, all those marrying for the first time after this age will be excluded. To avoid bias we must therefore ensure that age of marriage is an explanatory variable in the model and report results conditional on age of marriage. We consider first hazard-based models.

10.3   Hazard and survival functions

The underlying notions are those of *survivor* and *hazard* functions. Consider the (single-level) case where we have measures of length of employment on workers in a firm. We define the proportion of the workforce employed for periods greater than t as the *survivor function* and denote it by

$$S(t) = 1 - F(t) = 1 - \int_0^t f(u)du$$

where $f(u)$ is the density function of length of employment. The *hazard* function is defined as

$$h(t) = f(t)/S(t)$$

and represents the instantaneous risk, in effect the (conditional) probability of someone who is employed at time t, ending employment in the next (small) unit interval of time.

It is useful to carry out preliminary analyses of event history data by plotting the observed hazard and survivor functions. This can be done by dividing the time scale into short intervals and computing the empirical hazard for each interval by dividing the number who experience the event during the interval by the number of individuals present at the start of interval i, the 'risk set'. If there is censoring and we assume that the censoring is equally likely to have occurred at any time during the interval, then we subtract half the number of censored observations in that interval from the denominator. The survival function is simply the proportion who have not experienced the event by the start of interval i, plotted against time. Singer and Willett (2002) provide detailed examples.

The simplest model is one which specifies an exponential distribution for the duration time, $f(t) = \lambda e^{-\lambda t}(t \geq 0)$, which gives $h(t) = \lambda$, so that the hazard rate is constant and $S(t) = e^{-\lambda t}$. In general, however, the hazard rate will change over time and a number of alternative forms have been studied (see for example Cox and Oakes, 1984). A common one is based on the assumption of a Weibull distribution, namely

$$f(t) = (\alpha/t)e^{\alpha \ln(t)+\delta}e^{-e^{\alpha \ln(t)+\delta}}$$

or the associated extreme value distribution formed by replacing t by $u = \log e^t$.

Another approach to incorporating time-varying hazards is to divide the time scale into a number of discrete intervals within which the hazard rate is assumed constant, that is we assume a piecewise exponential distribution. This may be particularly useful where there are 'natural' units of time, for example based on menstrual cycles in the analysis of fertility, and this can be extended by classifying units by other factors where time varies over categories. We discuss such discrete time models in a later section.

The most widely used models are those known as proportional hazards models, and the most common definition is $h(t; \eta) = \lambda(t)e^\eta$. The term η denotes a linear function of explanatory variables, not functions of time, that we shall model explicitly in section 10.5. It is assumed that $\lambda(t)$, the baseline hazard function, depends only on time and that all other variation between units is incorporated into the linear predictor η. The components of η may also depend upon time, and in the multilevel case some of the coefficients will also be random variables.

10.4   Parametric proportional hazard models

For the case where we have known duration times and right censored data, define the cumulative baseline hazard function $\Lambda(t) = \int_0^t \lambda(u)du$ and a variable w with mean $\mu = \Lambda(t)e^\eta$, taking the value one for uncensored and zero for censored data. It can be shown (McCullagh and Nelder, 1989) that the maximum likelihood estimates required are those obtained from a maximum likelihood analysis for this model where w is treated as a Poisson variable. This computational device leads to the loglinear Poisson model for the i-th observation

$$\ln(\mu_i) = \ln(\Lambda(t_i)) + \eta_i \tag{10.1}$$

where the term $\Lambda(t_i)$ is treated as an offset, that is, a known function of the linear predictor.

The simplest case is the exponential distribution, for which we have $\Lambda(t) = \lambda t$. Equation (10.1) therefore has an offset $\ln(t_i)$ and the term $\ln(\lambda)$ is incorporated into η. We can model the response as a Poisson count using the procedures of Chapter 4, with coefficients in the linear predictor chosen to be random at levels 2 or above. This approach can be used with other distributions. For the Weibull distribution, of which the exponential is a special case, the proportional hazards model is equivalent to the log duration model with an extreme value distribution and we shall discuss its estimation in a later section.

10.5   The semiparametric Cox model

The most commonly used proportional hazard models are known as semiparametric proportional hazard models and we now look at the multilevel version of the most common of these in more detail.

Consider the 2-level proportional hazard model for the jk-th level 1 unit

$$h(t_{jk}; X_{jk}) = \lambda_k(t_{jk})\exp(X_{jk}\beta_k) \tag{10.2}$$

where X_{jk} is the row vector of explanatory variables for the level 1 unit and some or all of the β_k are random at level 2. We adopt the subscripts j, k for levels 1 and 2 for reasons which will be apparent below.

We suppose that the times at which a level 1 unit comes to the end of its duration period or 'fails' are ordered and at each of these we consider the total 'risk set'. At failure time t_{jk} the risk set consists of all the remaining level 1 units. Then the ratio of the hazard for the unit which experiences a failure and the sum of the hazards of the remaining risk set units is

$$\frac{\exp(X_{j'k'}\beta_{k'})}{\sum\limits_{j,k}\exp(X_{jk}\beta_k)}$$

which is simply the probability that the failed unit is the one denoted by j', k' (Cox, 1972). It is assumed that, conditional on the X_{jk}, these probabilities are independent.

Several procedures are available for estimating the parameters of this model (see for example Clayton, 1991, 1992). For our purposes it is convenient to adopt the following, which involves fitting a Poisson or equivalent multinomial model of the kind discussed in Chapter 7.

At each failure time l we define a response variate for each member of the risk set

$$y_{ijk(l)} = \begin{cases} 1 & \text{if } i \text{ is the observed failure} \\ 0 & \text{if not} \end{cases}$$

where i indexes the members of the risk set, and j, k level 1 and level 2 units. If we think of the basic 2-level model as one of employees within firms then we now have a 3-level model where each level 2 unit is a particular employee and containing n_{jk} level 1 units where n_{jk} is the number of risk sets to which the employee belongs. Level 3 is the firm. The explanatory variables can be defined at any level. In particular they can vary across failure times, allowing so called time-varying covariates. Overall proportionality, conditional on the random effects, can be obtained by ordering the failure times across the whole sample. In this case the *marginal* relationship between the hazard and the covariates generally is not proportional. Alternatively, we can consider the failure times ordered only *within* firms, so that the model yields proportional hazards within firms. In this case we can structure the data as consisting of firms at level 3, failure times at level 2 and employees within risk sets at level 1. In both cases, because we make the assumption of independence across failure times within firms, the Poisson variation is at level 1 and there is no variation at level 2. In other words we can collapse the model to two levels, within firms and between firms.

A simple variance components model for the expected Poisson count is written as

$$\pi_{jk(l)} = \exp(\alpha_l + X_{jk}\beta + u_k) \tag{10.3}$$

where there is a 'blocking factor' α_l for each failure time. In fact we do not need generally to fit all these nuisance parameters: instead we can obtain efficient estimates of the model parameters by modelling α_l as a smooth function of the time points, using, say, a low order polynomial or a spline function (Efron, 1988). Alternatively we can group the blocking factors into relatively homogeneous groups over longer time intervals, possibly based upon a preliminary inspection of the survival function.

For the model which assumes overall proportionality an estimator of the baseline surviving fraction for an individual in the k-th firm at time h, where $X_{jk} = 0$, is

$$\hat{S}_h = \exp\left(-\sum_{l \leq h} e^{\hat{\alpha}_l + \hat{u}_k}\right)$$

and the estimate for an individual with specific covariate values X_{jk} is

$$\hat{S}_h^{\{\exp(X_{jk}\beta)\}} \tag{10.4}$$

For the model which assumes proportionality within firms these two expressions become respectively

$$\hat{S}_h = \exp\left(-\sum_{l \leq h} e^{\hat{\alpha}_l}\right), \quad \hat{S}_h^{\{\exp(X_{jk}\beta) + \hat{u}_k\}}$$

where we fit polynomials to the blocking factors, the $\hat{\alpha}_l$ are estimated from the polynomial coefficients, and the surviving fraction can be plotted against the time associated with each interval.

10.6   Tied observations

We have assumed so far that each failure time is associated with a single failure. In practice many failures will often occur at the same time, within the accuracy of measurement. These are known as 'ties'. Sometimes, data may also be deliberately grouped in time. In this case all the failures at time l have a response $y_{ijk}(l) = 1$. This procedure for handling ties is equivalent to that described by Peto (1972; see also McCullagh and Nelder, 1989).

10.7   Repeated measures proportional hazard models

As in the case of ordinary repeated measures models described in Chapter 5 we can consider the case of multiple episodes, or durations within individuals, with between- and within-individual variation, and possibly further levels where individuals may be nested within firms, etc. The models of previous sections can be applied to such data, but there are further considerations which arise. Where each individual has the same fixed number n of episodes we can treat these, as in Chapter 6, as constituting n variates so that we have an n-variate model with an $(n \times n)$ covariance matrix between individuals. The variates may be either really distinct measurements or simply the different episodes in a fixed ordering. This is the model considered by Wei et al. (1989) who define proportionality as *within* individuals. We can also model a multivariate structure where, within individuals, there are repeated episodes for a number of different types of interval. For each type of interval we may have coefficients random at the individual level and these coefficients will generally also covary at that level.

Often with repeated measures models the first episode is different in nature from subsequent ones. An example might be the first episode of a disease which may tend to be longer or shorter than subsequent episodes. If the first episode is treated as if it were a separate variate then the subsequent episodes can be regarded as having the same distribution, as in the previous section. Another example is the length of the interval to a first birth from the start of a marriage or partnership. We shall give an example of such a model later.

Another possible complication in repeated measures data, as in Chapter 4, is that we may not be able to assume independence between durations within individuals. This will then lead to serial correlation models which can be estimated using the procedures discussed in Chapter 8 for the parametric log duration models discussed below. We shall return to a further exploration of repeated measures models when we look at discrete time models.

10.8   Example using birth interval data

The data are a series of repeated birth intervals for 379 Hutterite women living in North America (Egger, 1992; Larsen and Vaupel, 1993). The response is the length of time in months from birth to conception of next child, ranging from 1 to 160, with the first birth interval ignored and no censored information. This gives 2235 births in all.

There is information available on the mother's birth year, her age in years at the start of the birth interval, whether the previous child was alive or dead, and the duration of marriage at the start of the birth interval. Since we have a large number of women each with a relatively small number of intervals (although large compared to many studies with repeated measures) we have assumed overall proportionality, with failure times ordered across the whole sample. Table 10.1 gives the results for a variance components analysis and one where an additional random coefficient is estimated. A fourth order polynomial was adequate to smooth the blocking factors.

The only coefficient estimated with a non-zero variance at level 2 was whether or not the previous birth died, but a large sample chi-squared test, on 2 degrees of freedom, for the two random parameters for this coefficient gives a P-value of 0.01. An increase on the linear scale is associated with a shorter interval. Thus the birth interval decreases for the later born mothers and also if the previous child died. The interval is somewhat longer the longer the marriage duration with little additional effect of maternal age. This apparent lack of a substantial age effect seems to be a consequence of the high correlation (0.93) between duration of marriage and age. Higher order terms for duration and age were fitted but the estimated coefficients were small and not significant at the 10% level. The between-individual standard deviation is about 0.4 which is comparable in size to the effect of a previous death. The between-individual standard deviation for a model which fits no covariates is 0.45 so that the covariates explain only a small proportion of the between-individual variation. Figure 10.1 shows two average estimated surviving fraction curves for a woman aged 20, born in 1900 with

Table 10.1 Proportional hazards model for Hutterite birth intervals. In the random part subscript 0 refers to intercept, 1 to previous death

Parameter	Estimate (s.e.)	
	A	B
Fixed		
Intercept	−3.65	−3.64
Mother's birth year − 1900	0.026 (0.003)	0.026 (0.003)
Mother's age (year − 20)	−0.008 (0.014)	−0.004 (0.014)
Previous child died	0.520 (0.118)	0.645 (0.144)
Marriage duration (months)	−0.003 (0.001)	−0.004 (0.001)
Random		
σ^2_{u0}	0.188 (0.028)	0.188 (0.028)
σ_{u01}		0.005 (0.088)
σ^2_{u1}		0.381 (0.236)

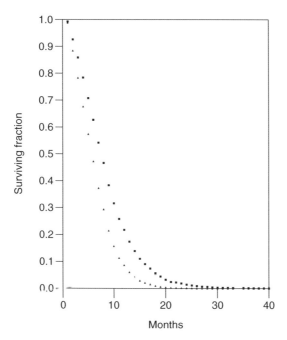

Figure 10.1 Probability of exceeding each birth interval length; live birth upper, previous death lower.

marriage duration 12 months. These curves represent the median, rather than mean, surviving fractions since we are here dealing with a nonlinear model. The higher one is for those where there was a previous live birth and the lower where there was a previous death.

10.9   Log duration models

For the accelerated life model the distribution function for duration is commonly assumed to be of the form

$$f(t; X, \beta) = f_0(te^{X\beta})e^{X\beta}$$

where f_0 is a baseline function (Cox and Oakes, 1984).

For a 2-level model this can be written as

$$l_{ij} = \log_e(t_{ij}) = X_{ij}\beta_j + e_{ij} \tag{10.5}$$

which is in the standard form for a 2-level model. We shall assume Normality for the random coefficients at level 2 (and higher levels) but at level 1 we shall study other distributional forms for the e_{ij}. The level 1 distributional form is important where there are censored observations. We first consider the common choice of an extreme value distribution for the log duration L, conditional on $X_{ij}\beta_j$, which as we noted above, implies an equivalence with the proportional hazards model. Omitting level subscripts we write

$$f(l; \alpha, \delta) = \alpha e^{-\alpha l + \delta} \exp(-e^{-\alpha l + \delta}) \quad -\infty < l < \infty \tag{10.6}$$

$$E(L) = \alpha^{-1}(\delta + \gamma), \quad \text{var}(L) = \frac{\pi^2}{6\alpha^2}, \quad \gamma = 0.5772$$

For (10.5) this gives

$$\pi_{ij} = \Pr(L > l_{ij}) = 1 - \exp(-e^{-\alpha l_{ij} + \delta_{ij}})$$

$$\pi'_{ij} = \alpha \cdot \exp\{-e^{-\alpha l_{ij} + \delta_{ij}}\}e^{-\alpha l_{ij} + \delta_{ij}} \tag{10.7}$$

where the differential is for use in the estimation of censored data and is with respect to β in the expression below.

The mean of L is incorporated into the fixed predictor. If we have no censored data we estimate the parameters for the model given by (10.5) by treating it as a standard multilevel model. We note that the estimation is strictly quasilikelihood since we are using only the mean and variance properties of the level 1 distribution. If we assume a simple level 1 variance then we can iteratively estimate α from the above relationship and we also obtain for the 2-level model (10.5)

$$\delta_{ij} = \gamma + \alpha(X_{ij}\beta_j)$$

Where there is complex variation at level 1 then α will vary with the level 1 units. To estimate the survival function for a given level 2 unit we first condition on the covariates and random coefficients, that is $X_{ij}\beta_j$, and then use (10.7).

We can choose other distributional forms for the log duration distribution. These include the log gamma distribution, the Normal and the logistic. Thus, for example, for the Normal distribution we have

$$\pi_{ij} = 1 - \Phi(z_{ij})$$
$$\pi'_{ij} = \phi(z_{ij})/\sigma_e$$
$$z_{ij} = [l_{ij} - (X\beta)_{ij}]/\sigma_e$$

where Φ, ϕ are the cumulative and density functions of the standard Normal distribution. Quasilikelihood estimates can be obtained for any suitable distribution with two parameters. The possibility of fitting complex variation at level 1 can be expected to provide sufficient flexibility using these distributions for most purposes.

10.9.1 Censored data

Where data are censored in log duration models we require the corresponding probabilities. Thus, for right censored data we would use (10.7) with corresponding formulae for interval or left censored data. For each censored observation we therefore have an associated probability, say π_{ij} with the response variable value of one.

This leads to a bivariate model, in which for each level 1 unit the response is either the continuous log duration time or takes the value one if censored with corresponding explanatory variables in each case. There are basically two sets of explanatory variables for the level 1 variation, one for the continuous log duration response and one for the binary response. In the former case we can extend this for complex level 1 variation, as in the example analysis below. For the latter we can use the standard logit model as described in Chapter 4, possibly allowing for extra-binomial variation. The random parameters at level 1 for the two components are uncorrelated. When carrying out the computations, we may obtain starting values for the parameters using just the uncensored observations.

Since the same linear function of the explanatory variables enters into both the linear and nonlinear parts of this model, we require only a single set of fixed part explanatory variables, although these will require the appropriate transformation for the logit response as described in Chapter 4. We also note that any kinds of censored data can be modelled, so long as the corresponding probabilities are correctly specified.

We can readily extend this model to the multivariate case where several kinds of durations are measured. This will require one extra lowest level to be inserted to describe the multivariate structure, with level 2 becoming the between-observation level and level 3 the original level 2. For the binary part of the model we will allow correlations at level 2 where these can be interpreted as point-biserial correlations. If we fit a probit model as discussed in Appendix 4.3, we can interpret such correlations on a continuous scale.

For repeated measures models where there are different types of duration we can choose to fit a multivariate model. Alternatively we may be able to specify a simpler model where the types differ only in terms of a fixed part contribution, or perhaps where there are different variances for each type with a common covariance. As pointed out earlier, we may sometimes wish to treat the first duration length separately and this is readily done by specifying it as a separate response.

10.9.2 Infinite durations

It is sometimes found that for a proportion of individuals, their duration lengths are extremely long or even infinite. Thus, some employees remain in the same job for life and some patients may acquire a disease and retain it for the rest of their lives. In studies of social mobility, some individuals will remain in a particular social group for a finite length of time while others may never leave it: such individuals we can refer to as 'long term survivors'. We will treat such durations as if they were infinite. Since any given study will last only for a finite time, it is impossible precisely to distinguish long term survivors from those which are right censored. Nevertheless, if we make suitable distributional assumptions we can obtain an estimate of the proportion of such survivors, θ.

For a constant θ, given an incompletely observed duration time, the observation is either right censored with finite duration or has infinite duration so that we replace the probability π_{ij} by $\lambda_{ij} = (1 - \theta)\pi_{ij} + \theta$. In general θ will depend on explanatory variables and an obvious choice for a model is

$$\text{logit}(\theta_{ij}) = X_{ij}^{(\theta)}\beta^{(\theta)} \tag{10.8}$$

Some of the coefficients in (10.8) may also vary across level 2 units.

Where the observation is not censored we know that it has a finite duration so that for an unobserved duration time we have a response variable taking the value zero with predictor given by $\{1 + \exp-(1 - \theta_{ij})\}^{-1}$. The full model can therefore be specified as a bivariate model where for observed durations we have two responses, one for the uncensored component l_{ij} and the one for the parameters $\beta^{(\theta)}$. For the incompletely observed observations there is a single response which takes the value one with predictor function

$$\{1 + \exp-[(1 - \theta_{ij})\pi_{ij} + \theta_{ij}]\}^{-1}$$

We can extend the procedures of Chapter 4 to the joint estimation of β, $\beta^{(\theta)}$, noting that for the censored observations when estimating β, we have

$$\lambda'_{ij}(\beta) = (1 - \theta_{ij})\pi'_{ij}$$

and for estimating β^θ we have

$$\lambda'_{ij}(\beta^{(\theta)}) = (1 - \pi_{ij})\theta'$$

10.10   Examples with birth interval data and children's activity episodes

We first look again at the Hutterite birth interval data. Since all the durations are uncensored we apply a standard model to the log(birth interval) values. Results are given in Table 10.2.

We see that we can now fit the year of birth and age as random coefficients at level 2. A joint test gives a chi-squared value of 10.4 with 5 d.f. P = 0.065, and they are each separately significant with a significance level of 6%. We have significant heterogeneity at level 1 where there is a greater variation in duration of intervals following a child death, with a chi squared on 1 d.f. of 4.7, P = 0.03. As before, mother's birth year and previous death are associated with a decrease and duration of marriage with an increase in birth interval. The estimated surviving fraction will in general depend on the level 1 distributional assumption.

In the present case, as shown in Figure 10.2, the level 1 standardized residuals show little departure from Normality and Figure 10.3 shows the estimated surviving fraction

Table 10.2   Log duration of birth interval for Hutterite women. Subscript 1 refers to birth year, 2 to age and 3 to previous death

Parameter	Estimate (s.e.)		
	A	B	C
Fixed			
Intercept	1.97	1.96	1.97
Mother's birth year − 1900	−0.021 (0.002)	−0.021 (0.002)	−0.021 (0.002)
Mother's age − 20	−0.005 (0.010)	−0.005 (0.010)	−0.005 (0.010)
Previous death	−0.435 (0.079)	0.436 (0.079)	−0.438 (0.089)
Marriage duration (months)	0.003 (0.001)	0.003 (0.001)	0.003 (0.001)
Random			
Level 2			
σ^2_{u0}	0.127 (0.017)	0.114 (0.052)	0.121 (0.054)
σ_{u01}		−0.001 (0.002)	−0.001 (0.002)
σ^2_{u1}		0.0001 (0.0001)	0.0001 (0.0001)
σ_{u02}		−0.004 (0.003)	−0.005 (0.003)
σ_{u12}		0.0001 (0.0001)	0.0001 (0.0001)
σ^2_{u2}		0.0005 (0.0003)	0.0006 (0.0003)
Level 1			
σ^2_{e0}	0.549 (0.018)	0.533(0.018)	0.522 (0.018)
σ^2_{e3}			0.200 (0.108)
−2 (log-likelihood)	5305.9	5295.5	5290.8

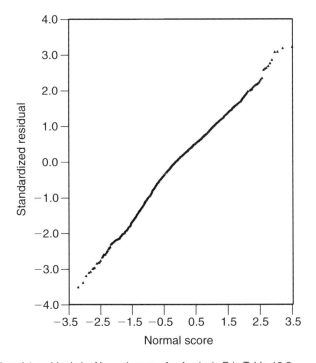

Figure 10.2   Level 1 residuals by Normal scores for Analysis B in Table 10.2.

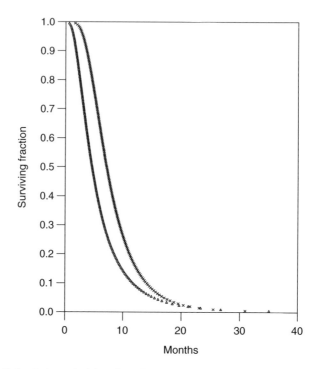

Figure 10.3 Estimated survival functions for women with previous live births (upper) and a previous death; born in 1900, age 20, 12 months' marriage.

based on Normality for women born in 1900, with marriage duration 12 months, aged 20 and with a previous survived birth.

Figure 10.3 is similar to Figure 10.1 based on the proportional hazards model. In fact, the two lines actually cross at about 30 months, as a result of the different level 1 variances for those with a previous survived birth as opposed to a death.

We now look at some data which exhibit more extensive variance heterogeneity at level 1. They measure the number of days spent by pre-school children either at home or in one of six different kinds of pre-school facility. For each of 249 children there were up to 12 periods of activity. Further details can be found in Plewis (1997; Chapter 7).

The response is the logarithm of the number of days and covariates are the type of episode, with home chosen as the base category and the education of the mother measured on a 7-point scale ranging from no education beyond minimum school leaving age (0) to university degree (6). Nineteen of the episodes were right censored and 25 were left censored, being less than one day.

The multilevel structure is that of episodes within children. The model is also multi-variate with the type of activity as six response variables, covarying at the level of the child. Table 10.3 shows the results of an analysis where there is a single between-child variance and where it is allowed to differ for each type of episode. The between-episode-within-child variance is also allowed to vary for different episodes. The level

Table 10.3 Log duration analysis of children's activity episodes: extreme value distribution

(a) Fixed parameter

	A (s.e.)	B (s.e.)
Fixed		
Intercept	2.19	2.10
Facility 1	−0.12 (0.11)	−0.13 (0.11)
Facility 2	0.20 (0.08)	0.18 (0.08)
Facility 3	0.00 (0.13)	0.00 (0.13)
Facility 4	0.87 (0.12)	0.95 (0.11)
Facility 5	0.28 (0.09)	0.28 (0.09)
Facility 6	0.15 (0.09)	0.14 (0.08)
Mother education	−0.05 (0.02)	−0.05 (0.02)

(b) Random parameter: level 1 variance

	A (s.e.)	B (s.e.)
Overall	0.75	
Home		0.76
Facility 1		1.23
Facility 2		0.83
Facility 3		0.79
Facility 4		0.40
Facility 5		0.65
Facility 6		0.57

(c) Level 2 covariance matrix: analysis A (analysis B in parentheses)

	Facility 1	*Facility 2*	*Facility 4*
Facility 1	0.34 (0.0)		
Facility 2	0.11 (0.0)	0.20 (0.17)	
Facility 4	−0.28 (0.0)	0.13 (0.09)	0.07 (0.23)

1 residuals for the continuous response part of the model show some evidence of non-Normality and we therefore show the results for the extreme value distribution. Because of the relatively small amount of censoring there is little difference for the parameter estimates between analyses making other distributional assumptions.

We see that there is quite substantial variation at both levels. At level 2 there was between-children variation only for facility types 1, 2 and 4. A proportional hazards model fitted to these data did not show any between-child variation. In general, the semiparametric proportional hazards model will not detect some of the relationships apparent from fitting parametric models, although it has the advantage that it does not make strong distributional assumptions. Figure 10.4 shows the predicted probabilities of home and facility type 1 episodes lasting beyond various times expressed in log (days). The crossing of the lines is now much clearer as a consequence of the different level 1 variances.

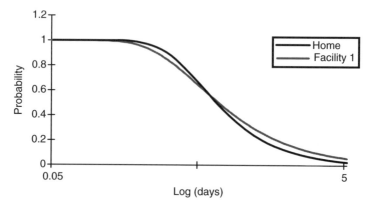

Figure 10.4 Estimated surviving probability of activity episodes.

10.11 The discrete time (piecewise) proportional hazards model

Where time is grouped into preassigned categories we write the survivor function at time interval t, the probability that failure occurs after this interval, as S_t. This gives

$$f_t = s_{t-1} - s_t, \quad h_t = f_t/s_{t-1}, \quad s_0 = 1$$

This gives

$$s_t = \prod_{l=1}^{t}(1 - h_l)$$

which can be used to estimate the survivor function from a set of estimated hazards. Thus the basic record is one record for each time interval within each higher level unit with the response being the proportion who fail during the interval. The estimation follows that for the binomial model described in Chapter 4 and we shall use the common choice of a logit link function, although other links are possible. For example, Aitkin et al. (1989) discuss a log-log link leading to a proportional hazards model. Censored observations are excluded from the relevant set.

As in the semi-parametric Cox model, we can fit a polynomial function to the successive time intervals, rather than the full set of blocking factors. The data will be ordered within level 2 units so that a risk set in general will extend over several such units. A general procedure is to specify the response for each level 1 unit as binary, that is zero if the unit survives the interval and one if not. Thus a 2-level model will become specified as a 3-level model with the binomial variation at level 1 and the actual level 1 units at level 2. This will be necessary if there are individual specific covariates. We now consider various extensions to this model to incorporate multivariate sequences, competing risks and multiple starting states. It is convenient to discuss these models for a 2-level repeated measures structure where each individual has a sequence of event episodes.

10.11.1 A 2-level repeated measures discrete time event history model

We shall discuss this model using an example from the National Child Development Study of co-residential partnership durations (Bynner et al., 2002). There are two

Table 10.4 Discrete time interval data for partnership duration

Individual	Actual time interval	Modelled time interval	Response	Event state
1	1	1	0	No partnership
1	2	2	0	No partnership
1	3	3	1	No partnership
I	4	1	0	Partnership
1	5	2	0	Partnership
1	6	3	0	Partnership
1	7	4	1	Partnership
1	8	1	0	No partnership
1	9	2	0	No partnership
1	10	3	0	No partnership

states – not being in a partnership (whether married or not) and belonging to a partnership. We assume that the total time interval is divided into short time intervals, for example 3 months, within which at most one transition is assumed to have taken place. These actual time intervals are re-coded into *modelled time intervals* (z) grouped within an episode (k), determined by the event state. For example data for individual 1 may look as in Table 10.4.

Thus, starting in state 'no partnership' (episode 1), individual 1 moves in time interval 3 to state 'partnership' (episode 2) and in time interval 7 to state 'no partnership' (episode 3). The response variable, y, takes the value zero if no move takes place and one if a change in partnership status occurs during the interval. Thus the hazard at modelled time t is

$$h_{ijk(t)} = P(y_{ijk(t)} = 1 | y_{ijk(t-1)} = 0)$$

where k indexes individual, j indexes episode and i indexes the state. The states (partnership, non-partnership) are modelled by dummy variables and this leads to a 3-level bivariate binary response model where, as we have assumed, the (conditional) responses within episodes within individuals are independent. Level 3 is the individual, level 2 represents variation between repeated episodes within individuals and level 1 refers to the time interval within repeated episodes.

Using a logit link function this model can be written in the form

$$\text{logit}(h_{ijk(t)}) = \beta_0 + \sum_{h=1}^{p} \alpha_h (z_{i(t)})^h + \sum_{l=1}^{m} \beta_l x_{lijk(t)} + u_{ijk} + v_{ik}$$

$$y_{ijk(t)} \sim \text{Bin}(1, h_{ijk(t)})$$

(10.9)

where z_{it} indexes for the modelled interval at discrete time t using a p-order polynomial (typically 4 or 5) to describe the baseline hazard (see section 10.5), and $x_{kij(t)}$ are covariates, including the dummy variable for partnership state. The term v_{ik} is the random effect for individual k for state i and u_{ijk} is the random effect associated with the j-th episode for the k-th individual. For simplicity we shall assume just a

2-level model with no within-individual-between-episode variation in the following exposition. Thus the model becomes

$$\text{logit}(h_{ijk(t)}) = \beta_0 + \sum_{h=1}^{p} \alpha_h (z_{i(t)})^h + \sum_{l=1}^{m} \beta_l x_{lijk(t)} + v_{ik} \qquad (10.10)$$

We retain the subscript j in the covariate expression to allow for episode varying covariates. Where there are many discrete time intervals this avoids the estimation of a large number of nuisance terms; one for each ordered time interval. At the individual level (2) we may have several random effects, in particular for each state, and these will covary.

The population probability of survival to the end of modelled time interval t for state i is

$$\prod_{m=1}^{t} (1 - h_{ijk(m)})$$

where these can be averaged over individual random effects to provide population estimates.

10.11.2 Partnership data example

The National Child Development Study (NCDS) is a longitudinal study which takes as its subjects all those living in Great Britain who were born between 3 and 9 March, 1958. The fifth follow-up took place when they were 33. A self-completion questionnaire asked for retrospective information on relationships, children, jobs and housing from the age of 16. Altogether, 11 178 persons responded and all but 39 of the cohort members had no more than four partnerships.

We confine ourselves to episodes that start with a partnership because the first (non-partnership) episode starting at age 16 is untypical (but see Goldstein et al., 2002b). At level 2 we have two random effects, one for partnership and one for non-partnership durations. The total number of 3-month periods is 14 0420 with 3737 male cohort members who had at least one partnership.

The MCMC estimation for (10.10) was run for 10 000 iterations with a burn in of 1000. It gives somewhat different estimates from PQL1, as we noted in Chapter 4. From Table 10.5 we see that there is some evidence that those with manual occupations have longer durations when in a partnership (the negative coefficient is associated with a lower probability of terminating an episode at any given time), but a small and non-significant difference ($-0.116 + 0.152 = 0.036$) for the non-partnership durations. The later the starting age the longer the duration for partnerships but there is only a small ($-0.063 + 0.049 = -0.014$) and non-significant relationship for non-partnerships. We note also that there is a negative correlation of -0.52 between partnership and non-partnership episodes, indicating that long partnerships tend to be associated with short non-partnerships and vice-versa. Thus individuals can, tentatively, be classified on this basis as either long partnership/short non-partnership or long non-partnership/short partnership individuals.

Table 10.5 Random coefficient model for partnership and outside partnership. MCMC estimates: starting from first partnership

Parameter	Estimate (s.e.)
Fixed	
Intercept	−3.709
z	$-0.244(0.067)^*10^{-1}$
z^2	$0.076(0.053)^*10^{-2}$
z^3	$-0.140(0.240)^*10^{-4}$
z^4	$-0.089(0.114)^*10^{-5}$
z^5	$0.093(0.264)^*10^{-7}$
Start age	−0.063(0.010)
Manual	−0.116(0.075)
np (non-partnership)	1.753(0.469)
np^*z	$0.499(0.208)^*10^{-1}$
np^*z^2	$-0.181(0.134)^*10^{-2}$
np^*z^3	$-0.112(0.114)^*10^{-3}$
np^*z^4	$0.024(0.332)^*10^{-5}$
np^*z^5	$0.055(0.158)^*10^{-6}$
np^* start age	0.049(0.017)
np^* manual	0.152(0.118)
Random	
σ_{v0}^2 (non-partnership)	0.400(0.211)
σ_{v01}	−0.119(0.118)
σ_{v1}^2 (partnership)	1.145(0.171)

10.11.3 General discrete time event history models

We now look at some general extensions to the model. We write (10.9) in the general form for a 2-level model

$$\text{logit}(h_{ij(t)}) = (Z\alpha)_{it} + (X\beta)_{ij(t)} + u_{ij}$$
$$y_{ij(t)} \sim Bin(1, h_{ij(t)}), \quad u_{ij} \sim MVN(0, \Omega_u) \tag{10.11}$$

Equation (10.11) is a proportional hazards model but non-proportional hazards can be introduced by including interactions between the Z and X variables. We also note that it allows time varying covariates. In general we can model a cohort of individuals moving through time where the response is the proportion who fail during an interval. For simplicity we assume in the following discussion that the observations are for intervals within individuals, as in the example of the previous section.

A *competing risks* model is one where there are several ways for an interval to end, in addition to several ways for states to start, indexed by i. Suppose that there are $R - 1$ end events. Denote the multinomial response by $y_{ij(t)}$ where $y_{ij(t)} = r$ if an event of type r has occurred in time interval $t, r = 2, \ldots, R$, and $y_{ij(t)} = 1$ if no event has occurred. The hazard of an event of type r in interval t, denoted by $h_{ij(t)}^{(r)}$, is the probability that an event of type r occurs in interval t, for state i given that no event of any type has occurred before interval t. The log-odds of an event of type r versus no event is modelled using a multinomial logit model as defined in Chapter 4 as follows

$$\log\left(\frac{h_{ij(t)}^{(r)}}{h_{ij(t)}^{(1)}}\right) = (Z\alpha)_{(t)}^{(r)} + (X\beta)_{ij(t)}^{(r)} + u_{ij}^{(r)}, \qquad r = 2, \ldots, R \qquad (10.12)$$

We may also have different sets of end events for each state in which case (10.12) is modified so that there are R_i end events for state i. In a competing risks model, the effects of duration and covariates may differ for each event type, as indicated by the r superscript. It is also possible that the form of the baseline hazard and the set of covariates may vary across event types. The random effects are assumed to follow a multivariate normal distribution, with covariance matrix Ω_u and we will in general have an effect for each combination of state and end event. A simplification will occur if we are prepared to assume that, for each end event, there is a random effect irrespective of initial state so that (10.12) becomes

$$\log\left(\frac{h_{ij(t)}^{(r)}}{h_{ij(t)}^{(1)}}\right) = (Z\alpha)_{(t)}^{(r)} + (X\beta)_{ij(t)}^{(r)} + u_i^{(r)} + u_j^{(r)} \qquad (10.13)$$

which has the form of a cross-classified model (see Chapter 11).

We can fit all of these models using alternative link functions such as the probit or complementary log-log. In the case of the probit link we can interpret the hazard as the cumulative probability for an underlying standard Normal distribution. Thus, for a given time period, covariates (and random effects) corresponding to an observed response, we can estimate the value on the underlying Normal density, interpreted as a propensity to end an episode at time t (see Chapter 4). The values of these at any time and set of covariate values therefore provides an alternative interpretation for the model.

Where the transition states form an ordered categorization we can use corresponding ordered category models for this, for example by modelling cumulative log-odds (see Chapter 4). This could arise, for example, in the modelling of illness duration where patients make transitions between clinical states which are ordered by severity. For such models we can also assume an underlying propensity with a probit link and this can be fitted via MCMC.

These models can also be extended readily to the multivariate case where, for each individual, we wish to study more than one type of episode at a time; for example duration of contraceptive use and duration of employment episodes. For each episode type we form the same set of discrete elementary time intervals and the response is a p-way table where p is the number of episode types. This table is treated as a multinomial response with corresponding, dummy, explanatory variables for the margins of the table that we wish to fit. If covariates are present these will be modelled by interacting them with the explanatory variable dummies. For ordered models and for binary response models with a probit link function, we can directly incorporate correlations between the underlying Normal distributions at level 1. This then provides covariance matrix estimates for the episode types at all levels of the data hierarchy.

Discrete time models provide a very flexible means of modelling multilevel event history data since they just involve the use of existing procedures for discrete responses. One disadvantage is that they require data in a form that uses large amounts of memory, but that is likely to become less important with advances in computer technology. Steele et al. (2003) discuss these models in more detail with examples, and show how they can be fitted with existing software.

11
Cross-Classified Data Structures

11.1 Random cross-classifications

In previous chapters we have considered only data where the units have a purely hierarchical or nested structure. In many cases, however, a unit may be classified along more than one dimension. An example is students classified both by the school they attend and by the neighbourhood where they live. We can represent this diagrammatically as in Figure 11.1 for three schools and four neighbourhoods with between one and six students per school/neighbourhood cell. The cross-classification is at level 2 with students at level 1.

Another example is in a repeated measures study where children are measured by different raters at different occasions with up to two measurements at each occasion. If each child has its own set of raters not shared with other children then the cross-classification is of occasions and raters within children. This can be represented diagrammatically in Figure 11.2 for three children with up to seven measurement occasions and two raters per child. If we adopt the convention that the lowest level at which a response occurs is level 1, then the cross-classification here is at level 2 with level 3 that of the child. This is a special case of a level 2 cross-classification with only one unit per cell (see also section 11.6).

If now the same set of raters is involved with all the children the crossing is at the child level, 2, as can be seen in Figure 11.3 with three raters and three children and up to five occasions. Figure 11.3 is formally the same structure as Figure 11.1 but with the level 1 variance being that between occasions,

These basic cross classifications occur commonly when a simple hierarchical structure breaks down. Consider, for example, a repeated measures design which follows a

	School 1	School 2	School 3
Neighbourhood 1	x x x x	x x	x
Neighbourhood 2	x	x x x x x x	x x x
Neighbourhood 3	x x	x	x x x x
Neighbourhood 4	x x x	x x	x x

Figure 11.1 A random cross-classification at level 2.

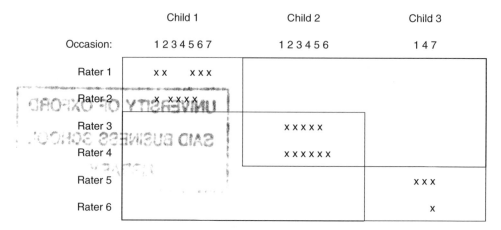

Figure 11.2 A random cross-classification at level 2 with one unit per cell.

	Child 1 1 2 3 4	Child 2 1 2	Child 3 1 2 3 4 5
Rater 1	x x x	x	x
Rater 2	x		x x
Rater 3		x	x x

Occasion: (Child 1) 1 2 3 4 (Child 2) 1 2 (Child 3) 1 2 3 4 5

Figure 11.3 A random cross-classification at level 2.

sample of students over time, say once a year, within a set of classes for a single school. We assume first that each class group is taken by the same teacher. The hierarchical structure is then a 3-level one with occasions grouped within students who are grouped within classes. If we had several schools then schools would constitute the level 4 units. Suppose, however, that students change classes during the course of the study. For three students, three classes and up to three occasions we might have the pattern shown in Figure 11.4.

Formally this is the same structure as Figure 11.3, that is a cross-classification at level 2 for classes by students. Such designs will occur also in panel or longitudinal studies of individuals who move from one locality to another, or workers who change their place of employment. If we now include schools these will be classified as level 3 units, but if students also change schools during the course of the study then we obtain a level 3 cross-classification of students by schools with classes nested at level 2 within schools and occasions as the level 1 units. The students have moved from being crossed with classes to being crossed with schools. Note that since students are crossed at level 3 with schools they are also automatically crossed with any units nested within schools so that we do not need separately to specify the crossing of classes with students.

Suppose now that, instead of the same teachers taking the classes throughout the study, the classes are taken by a completely new set of teachers every year and where

	Student 1	Student 2	Student 3
Occasion:	1 2 3	1 2	1 2 3
Class/teacher 1	x x	x	x
Class/teacher 2	x		
Class/teacher 3		x	x x

Figure 11.4   Students changing classes/teachers.

		Student 1	Student 2	Student 3
	Year:	1 2	1 2	1 2
Teacher 1	1	x	x	
Teacher 2				x
Teacher 3	2	x		x
Teacher 4			x	

Figure 11.5   Students changing teachers and groups.

new groupings of students are formed each year too. Such a structure with four different teachers over 2 years for three students is given in Figure 11.5.

This is now a cross-classification of teachers by students at level 2 with occasion as the level 1 unit. We note that most of the cells are empty and that there is at most one level 1 unit per cell so that no independent between-occasion variance can be estimated as pointed out above. Raudenbush (1993) gives an example of such a design, and provides details of an EM estimation procedure for 2-level two-way cross-classifications with worked examples.

We can have a design which is a mixture of those given by Figures 11.4 and 11.5 where some teachers are retained and some are new at each occasion. In this case we would have a cross-classification of teachers by students at level 2 where some of the teachers only had observations at one occasion. More generally, we can have an unbalanced design where each teacher is present on a variable number of occasions. With two occasions where we have the same teachers or intact groups we can formulate an alternative cross-classification design which may be more appropriate in some cases. Instead of cross-classifying students by teachers we consider cross-classifying the set of all teachers at the first occasion by the same set at the second occasion, as follows.

Consider Figure 11.6 where we have 22 students who are nested within the cross-classification of teachers at each occasion. The difference between this design and that in Figure 11.4 is analogous to the difference between a two-occasion longitudinal design where a second occasion measurement is regressed on a first occasion measurement and the two-occasion repeated measures design where a measurement is related to age or time. In Figure 11.6 we are concerned with the contribution from each occasion to the variation in, say, a test score measured at occasion 2, and we might consider

Occasion 2

	Teacher 1	Teacher 2	Teacher 3
Teacher 1	x x x x x	x	x x
Teacher 2	x x	x x x x	
Teacher 3	x	x x x	x x x x

Occasion 1 (row labels: Teacher 1, Teacher 2, Teacher 3)

Figure 11.6 Teachers cross-classified by themselves at two occasions with responses from 22 students.

any first occasion measurement as an explanatory variable. In Figure 11.4 on the other hand, although we could fit a separate between-teacher variance for each occasion, the response variable is essentially the same one measured at each occasion.

In all the cases of longitudinal data we have considered where students move between teachers or schools the response variable is measured at each occasion. In cases where a single, for example final occasion, response occurs we have a model where the level 2 (and above) contributions are averaged over all the units which the student attends. In such cases we require a *multiple membership* model and this is described in Chapter 12.

11.2   A basic cross-classified model

Goldstein (1987a) sets out the general structure of a model with both hierarchical and cross-classified structures. We consider first the simple model of Figure 11.1 with variance components at level 2 and a single variance term at level 1.

We shall refer to the two classifications at level 2 using the subscripts j_1, j_2 and in general parentheses will group classifications at the same level. We write the model as

$$y_{i(j_1 j_2)} = X_{i(j_1 j_2)}\beta + u_{1j_1} + u_{2j_2} + e_{i(j_1 j_2)} \qquad (11.1)$$

The covariance structure at level 2 can be written in the following form

$$\text{cov}(y_{i(j_1 j_2)} y_{i'(j_1 j_2')}) = \sigma_{u_1}^2$$
$$\text{cov}(y_{i(j_1 j_2)} Y_{i'(j_1' j_2)}) = \sigma_{u_2}^2 \qquad (11.2)$$
$$\text{var}(y_{i(j_1 j_2)}) = \text{cov}(y_{i(j_1 j_2)} y_{i'(j_1 j_2)}) = \sigma_{u_1}^2 + \sigma_{u_2}^2$$

We note that if there is no more than one level 1 unit per cell model (11.1) is still valid.

The level 2 variance is the sum of the separate classification variances, the covariance for two level 1 units in the same classification is equal to the variance for that classification and the covariance for two level 1 units which do not share either classification is zero. If we have a model where random coefficients are included for either or both classifications, then analogous structures are obtained. We can also add further ways of classification with obvious extensions to the covariance structure.

Appendix 11.1 shows how cross-classified models can be specified and estimated efficiently, using a purely hierarchical formulation with the IGLS algorithm. For large

Table 11.1 Analysis of examination scores by secondary and by primary school attended. The subscript 1 refers to primary and 2 to secondary school

Parameter	Estimate (s.e.)		
	A	B	C
Fixed			
Intercept	5.50	5.98	5.99
Verbal reasoning	—	0.16 (0.003)	0.16 (0.003)
Random			
$\sigma_{u_1}^2$	1.12 (0.20)	0.27 (0.06)	—
$\sigma_{u_2}^2$	0.35 (0.16)	0.011 (0.021)	0.28 (0.06)
σ_e^2	8.1 (0.2)	4.25 (0.10)	4.26 (0.10)

problems, however, that is where two or more cross-classifications contain large numbers of units, this procedure becomes unwieldy and in a later section we shall show how MCMC estimation can handle such situations and we will also introduce a more general notation.

11.3 Examination results for a cross-classification of schools

The data consist of scores on school leaving examinations obtained by 3435 students who attended 19 secondary schools cross-classified by 148 primary schools in Fife, Scotland (Paterson, 1991). Before their transfer to secondary school at the age of 12 each student obtained a score on a verbal reasoning test, measured about the population mean of 100 and with a population standard deviation of 15.

The model is as follows

$$y_{i(j_1 j_2)} = \beta_0 + \beta_1 x_{1i(j_1 j_2)} + u_{1j_1} + u_{2j_2} + e_{i(j_1 j_2)} \tag{11.3}$$

$$u_{1j_1} \sim N(0, \sigma_{u_1}^2), \quad u_{2j_2} \sim N(0, \sigma_{u_2}^2), \quad e_{i(j_1 j_2)} \sim N(0, \sigma_e^2)$$

and the results are given in Table 11.1. Random coefficients for verbal reasoning were also fitted but the associated variances are estimated as zero.

Ignoring the verbal reasoning score (model A) we see that the between-primary school variance is estimated to be more than three times that between secondary schools. One reason for this may be that the secondary schools are on average far larger than primary schools, so that the sampling variance is smaller. Such an effect will often be observed where one classification has far fewer units than another, for example where a small number of schools is crossed with a large number of small neighbourhoods or a small number of teachers is crossed with a large number of students at level 1 within schools. In such circumstances we need to be careful about our interpretation of the relative sizes of the variances. To study further the issue of school size we could make the between-school variance a function of school size, but this variable is not available in the present dataset.

When the verbal reasoning score (model B) is added to the fixed part of the model the between-secondary school variance becomes very small, the between-primary school variance is also considerably reduced and the level 1 variance also. The third analysis (model C) shows the effect of removing the cross-classification by primary school. The between-secondary school variance is now only a little smaller than in the analysis A without verbal reasoning score. Using analysis C alone, which is typically the case with school effectiveness studies which control for initial achievement, we would conclude that there were important differences between the progress made in secondary schools. From analysis B, however, we see that most of this is explained by the primary schools attended. Of course, the verbal reasoning score is only one measure of initial achievement, but these results illustrate that adjusting for achievement at a single previous time may not be adequate.

11.4   Interactions in cross-classifications

Consider the following extension of equation (11.1)

$$y_{i(j_1 j_2)} = X_{i(j_1 j_2)}\beta + u_{1j_1} + u_{2j_2} + u_{3(j_1 j_2)} + e_{i(j_1 j_2)} \tag{11.4}$$

We have now added an 'interaction' term to the model which was previously an additive one for the two random effects. The usual specification for such a random interaction term is that it has a simple variance $\sigma^2_{u_{(12)}}$ across all the level 2 cells (Searle et al., 1992). The adequacy of such a model can be tested against an additive model using a likelihood ratio test criterion. For the example in Table 11.1 this interaction term is estimated as zero. While this indicates that the cross-classification is adequate, because the between-secondary school variance is so small we would not expect to be able to detect such an interaction.

Extensions to this model are possible by adding random coefficients for the interaction component, just as random coefficients can be added to the additive components. For example, the gender difference between students may vary across both primary and secondary schools in the example of section 11.3 and we can fit an extra variance and covariance term for this to both the additive effects and the interaction.

11.5   Cross-classifications with one unit per cell

Some interesting models occur when there is only one level 1 unit per cell of a level 2 cross-classification. This case should be distinguished from the case where a level 2 cross-classification happens to produce no more than one level 1 unit in a cell as a result of sampling. Thus (11.4), where there are some cells with more than one level 1 unit, allows us to obtain separate estimates for the level 1 variance and the level 2 interaction. If there is only one level 1 unit per cell then this interaction is confounded with the level 1 variance.

So called 'generalizability theory' models (Cronbach and Webb, 1975) can be formulated as a cross-classification with one level 1 unit per cell. The basic model is one where a test or other instrument consisting of a set of items, for example ratings or questions, is administered to a sample of individuals. The individuals are therefore cross-classified by the items and may be further nested within schools, etc. at higher levels. In educational test settings the item responses are often binary so that we would apply the methods of Chapter 4 to the present procedures in a straightforward way. Since each individual can only respond once to each item this is an example where we cannot directly detect a level 2 interaction.

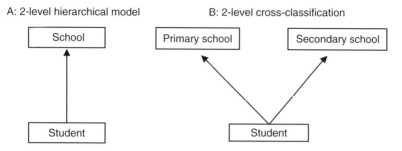

Figure 11.7 Classification diagrams for hierarchical and cross-classified structures

11.6 Multivariate cross-classified models

For multivariate models the responses may have different structures. Thus in a bivariate model one response may have a 2-level hierarchical structure and the other may have a cross-classification at level 2. Suppose, for example, that we measure the height and the mathematics attainment of a sample of students from a sample of schools. The mathematics attainment is assessed by a different set of teachers in each school and the heights are measured by a single anthropometrist. For the mathematics scores there is a cross-classification of students by teachers within each school whereas for height there is a 2-level hierarchy with students nested within schools. Height and mathematics attainment will be correlated at both the student and the school level and we can write a model for this structure as follows

$$y_{h(i_1 i_2)j} = \delta_{1h}(X_{1(i_1 i_2)j}\beta_1 + u_{1j} + e_{1i_1 j} + e_{1i_2 j}) + \delta_{2h}(X_{2i_1 j}\beta_2 + u_{2j} + e_{2i_1 j})$$
$$\text{cov}(u_{1j}u_{2j}) = \sigma_{u12} \quad \text{cov}(e_{1i_1 j}e_{2i_1 j}) = \sigma_{e12} \tag{11.5}$$
$$\delta_{1h} = 1 \ \ if \ mathematics, \quad 0 \ \ if \ height, \quad \delta_{2h} = 1 - \delta_{1h}$$

where all other covariances are zero.

We have already mentioned that cross-classified models can have a discrete response. We can also fit, for example, time series models as discussed in Chapter 5 and in general cross-classified structures can incorporate all the types of models which can be fitted for purely hierarchical structures.

11.7 A general notation for cross-classifications

In Figure 11.7 we set out a simple diagram, called a 'classification diagram' and introduced by Browne et al. (2001b). It allows us to classify data structures as hierarchical or crossed or combinations of these at different levels. Boxes represent unit classifiers and those at the same horizontal level in a cross-classification are at the same level conceptually. In terms of the model we also have a simplified notation; for a basic variance components model we write

$$y_i^{(1)} = (X\beta)_i + u_{school(i)}^{(2)} + u_{student(i)}^{(1)} \ \ school(i) \in (1,\ldots.J) \ \ student(i) \in (1,\ldots.N)$$
$$u_{school(i)}^{(2)} \sim N(0,\sigma_{u(2)}^2) \quad u_{student(i)}^{(1)} \sim N(0,\sigma_{u(1)}^2) \quad i = 1,\ldots.N \tag{11.6}$$

In (11.6) we now have two *classifications*, with students as classification 1 and schools as classification 2. The subscript i is attached to the lowest level units and uniquely identifies every measurement and random effect. Thus *school(i)* is the school that student i belongs to and we refer to *school(i)* as a classification function that maps the lowest level units, students, onto schools. *Student(i)* is the identity function that maps students onto themselves. The superscript denotes the classification, where the lowest is numbered 1, etc. Together with diagram A in Figure 11.7 equation (11.6) completely specifies the model.

We can now rewrite (11.1) as

$$y_i^{(1)} = (X\beta)_i + u_{neighbourhood(i)}^{(3)} + u_{school(i)}^{(2)} + u_{student(i)}^{(1)} \qquad (11.7)$$

$$neighbourhood\,(i) \in (1,\ldots.J_3), \quad school(i) \in (1,\ldots.J_2),$$

$$student(i) \in (1,\ldots.N)$$

$$u_{neighbourhood(i)}^{(3)} \sim N(0,\sigma_{u(3)}^2),\; u_{school(i)}^{(2)} \sim N(0,\sigma_{u(2)}^2),$$

$$u_{student(i)}^{(1)} \sim N(0,\sigma_{u(1)}^2) \quad i = 1,\ldots.N$$

which together with diagram B in Figure 11.7 completely specifies the cross-classified model. Since the only subscript is that for the lowest level units the notation can be extended indefinitely for any number of crossed or hierarchical classifications. The earlier notation as in (11.1) becomes very cumbersome when many classifications exist. Browne et al. (2001b) give a comprehensive treatment of this general notation which, as we shall see in Chapter 12, extends readily to handle multiple membership data structures. These authors also adopt the convention of dropping the (1) superscript for level 1 responses and effects. This provides somewhat greater clarity for models such as (11.7) but there are models, for example with multivariate responses at several levels, where it is useful to retain the level 1 superscript.

11.8   MCMC estimation in cross-classified models

The extension of MCMC methods to cross-classifications is straightforward since we simply introduce further steps for each extra classification that samples residuals and covariance matrices from the relevant conditional distribution. Thus, for example for (11.9), the Gibbs algorithm in section 2.13 becomes

- **Step 1**
  Sample a new set of fixed effects (β).
- **Step 2**
  Sample a new set of neighbourhood residuals $\{u^{(3)}\}$.
- **Step 3**
  Sample a new set of school residuals $\{u^{(2)}\}$.
- **Step 4**
  Sample a new neighbourhood classification variance.
- **Step 5**
  Sample a new school classification variance.
- **Step 6**
  Sample a new level 1 variance.

- **Step 7**

  Compute the level 1 residuals by subtraction.

Similar modifications to MCMC algorithms allow cross-classified models with discrete responses, etc. For starting values typically it will be possible to use estimates obtained from fitting a purely hierarchical model for each single higher level classification in turn. If we use MCMC with default Gamma priors to fit the Fife examination data we obtain results very similar to those in Table 11.1.

Appendix 11.1

IGLS estimation for cross-classified data

11.1.1 An efficient IGLS algorithm

We illustrate the procedure using a 2-level model with crossing at level 2. The 2-level cross-classified model, using the notation in Appendix 2.1, can be written as

$$y_{i(j_1 j_2)} = X_{i(j_1 j_2)}\beta + \sum_{h=1}^{q_1} z_{1hij_1} u_{1hj_1}$$

$$+ \sum_{h=1}^{q_2} z_{2hij_2} u_{2hj_2} + e_{i(j_1 j_2)} \tag{11.1.1}$$

Parentheses group the ways of classification at each level. We have two sets of explanatory variables, type 1 and type 2, for the random components defined by the columns of $Z_1(n \times p_1 q_1)$, $Z_2(n \times p_2 q_2)$ where p_1, p_2 are respectively the number of categories of each classification.

$$Z_1 = \{z_{1hij_1}\}, \quad Z_2 = \{z_{2hij_2}\}$$

$$z_{1hij_1} = z_{1him} \quad \textit{if } j_1 = m, \textit{ for m-th type 1 level 2 unit}, \quad 0 \textit{ otherwise}$$

$$z_{2hij_2} = z_{2him} \quad \textit{if } j_2 = m, \textit{ for m-th type 2 level 2 unit}, \quad 0 \textit{ otherwise}$$

These variables are dummy variables where for each level 2 unit of type 1 we have q_1 random coefficients with covariance matrix $\Omega_{(1)2}$ and likewise for the type 2 units. To simplify the exposition we restrict ourselves to the variance component case where we have

$$\Omega_{(1)2} = \sigma_{(1)2}^2, \quad \Omega_{(2)2} = \sigma_{(2)2}^2$$

$$E(\tilde{Y}\tilde{Y}^T) = V_1 + Z_1(\sigma_{(1)2}^2 I_{(p_1)})Z_1^T + Z_2(\sigma_{(2)2}^2 I_{(p_2)})Z_2^T \tag{11.1.2}$$

Consider Figure 11.1 in Chapter 11 where schools are ordered within neighbourhoods. The explanatory variables will have the structure shown in Figure 11.1.1 for the first eight students.

It is clear that the second term in (11.1.2) can be written as

$$Z_1(\sigma_{(1)2}^2 I_{(p_1)})Z_1^T = J\sigma_{(1)2}^2 J^T$$

where J is a $(n \times 1)$ vector of ones. The third term is of the general form $Z_3\Omega_3 Z_3^T$, namely a level 3 contribution where in this case there is only a single level 3 unit and

i,j_1	i,j_2	Z_{11}	Z_{12}	Z_{13}	Z_{14}	Z_{21}	Z_{22}	Z_{23}
1,1	1,1	1	0	0	0	1	0	0
2,1	2,1	1	0	0	0	1	0	0
3,1	3,1	1	0	0	0	1	0	0
4,1	4,1	1	0	0	0	1	0	0
5,1	1,2	1	0	0	0	0	1	0
6,1	2,2	1	0	0	0	0	1	0
7,1	1,3	1	0	0	0	0	0	1
1,2	2,1	0	1	0	0	1	0	0

Figure 11.1.1   Explanatory variables for level 2 cross-classification of Figure 11.1.

with no covariances between the random coefficients of the Z_{2h} and with the variance terms constrained to be equal to a single value, $\sigma^2_{(2)2}$.

More generally we can specify a level 2 cross-classified variance components model by modelling one of the classifications as a standard hierarchical component and the second as a set of dummy explanatory variables, one for each category, with the random coefficients uncorrelated and with variances constrained to be equal. If this second (type 2) classification has further explanatory variables with random coefficients as in (11.1.1), then we form extended dummy variable 'interactions' as the product of the basic dummy variables and the further explanatory variables with random coefficients, so that these coefficients have variances and covariances within the same type 2 level 2 unit but not across units. In addition the corresponding variances and covariances are constrained to be equal.

To extend this to further ways of classification we add levels. Thus, for a three-way cross-classification at level 2 we choose one classification, typically that with the largest number of categories, to model in standard hierarchical fashion at level 2, the second to model with coefficients random at level 3 as above and the third to model in a similar fashion with coefficients random at level 4. We can also allow simultaneous crossing at more than one level. Thus for example, if there is a 2-way cross-classification at level 2 and a three-way cross-classification at level 3, we will require five levels, the first two describing the level 2 cross-classification and the next three describing the level 3 cross-classification. Further details are given by Rasbash and Goldstein (1994).

11.1.2   Computational considerations

Analysis A in Table 11.1 took about 40 seconds per iteration on a 66 Mhz 486 PC using the MLwiN software, approximately ten times longer than analysis C. This relative slowness is due to the size of the single level 3 unit which contains all the 3435 level 1 units. For very much larger problems the computing considerations will become of greater concern, so that some procedure for speeding up the computations is needed.

Table 11.1.1 Examination scores for secondary by primary school classification omitting small cells

Parameter	Estimate (s.e.)	
	≤1 student	≤2 students
Fixed		
Intercept	6.00	6.00
Verbal reasoning	0.16 (0.003)	0.16 (0.003)
Random		
$\sigma^2_{u_{(1)0}}$	0.27 (0.06)	0.25 (0.06)
$\sigma^2_{u_{(2)0}}$	0.004 (0.021)	0.028 (0.030)
σ^2_e	4.28 (0.11)	4.29 (0.11)

In the present analysis there are 120 cells of the cross-classification which contain only one student. If we eliminate these data from the analysis we obtain two disjoint subsets containing 14 and five secondary schools. There are a further 24 cells containing two students and if these are removed we obtain six disjoint subsets the largest of which contains eight secondary schools. Table 11.1.1 shows the estimates from the resulting analyses.

The only substantial difference is in the between-secondary school variance which is anyway poorly estimated. The first analysis took about 15 seconds and the second about 6 seconds. In many cases such separations may not be possible and MCMC estimation (see section 11.8) becomes relatively fast and the preferred method. It also has the advantage that it is unnecessary to eliminate any data as has been done here.

12

Multiple Membership Models

12.1   Multiple membership structures

In some circumstances units can be members of more than one higher level unit at the same time. An example is friendship patterns where at any time individuals can be members of more than one friendship group. In an educational system students may attend more than one institution. In all such cases we shall assume that for each higher level unit to which a lower level unit belongs there is a known weight (summing to 1.0 for each lower level unit), which represents, for example, the amount of time spent in that unit. The choice of weights may be important and in practice a sensitivity analysis will often be useful to determine how alternative choices of weights affect inferences.

We may also have data where there is some uncertainty about which higher level unit some lower level units belong to. For example, in a survey of students, information about their neighbourhood of residence may only be available for a few students for larger geographical units. For these cases it may be possible to assign a weight for each of the constituent neighbourhoods which is in effect a probability of belonging to each based upon available information. We will refer to such a structure as a missing identification structure for which the multiple membership estimation techniques are readily adapted. Hill and Goldstein (1998) discuss both multiple membership and missing identification models with examples.

Consider the 2-level variance components model with each level 1 unit belonging to at most two level 2 units where the j_1, j_2 subscripts now refer to the same type of unit, so that we do not require separate subscripts $(1, 2, 3, \ldots)$ for the u_j terms.

$$y_{i(j_1 j_2)} = X_{i(j_1 j_2)}\beta + w_{1ij_1}u_{j_1} + w_{2ij_2}u_{j_2} + e_{i(j_1 j_2)}$$
$$w_{1ij_1} + w_{2ij_2} = 1 \tag{12.1}$$

The overall contribution at level 2 is the weighted sum over the level 2 units to which each level 1 unit belongs. This leads to the following covariance structure

$$\text{var}(y_{i(j_1 j_2)}) = (w_{1ij_1}^2 + w_{2ij_2}^2)\sigma_u^2 + \sigma_e^2$$
$$\text{cov}(y_{i(j_1 j_2)}y_{i'(j_1 j_2)}) = (w_{1ij_1}w_{1i'j_1} + w_{2ij_2}w_{2i'j_2})\sigma_u^2$$
$$\text{cov}(y_{i(j_1 j_2)}y_{i'(j_1' j_2)}) = w_{2ij_2}w_{2i'j_2}\sigma_u^2$$

For individuals belonging to more than one unit, the contribution to the total variation of the higher level units is less; e.g.

$$\text{If } w_{i1} = w_{i2} = 0.5, \quad \text{var}\left(\sum_h w_{ih} u_h\right) = \sigma_u^2/2 \tag{12.2}$$

This has important implications for models which ignore large numbers of multiple memberships. For example, suppose in a study of school examination results, a substantial number of students have spent time in several schools but they are assigned only to the final school where they took the examination. The observed level 2 variation will be less than the true level 2 variation for these students as demonstrated in (12.2). The effect will be to underestimate the true level 2 variation and thus to produce biased estimates for school-level residuals. It would seem that almost all so-called 'school effectiveness' studies, including those used for illustration in earlier chapters, are subject to such biases, although it is not clear how important these are.

12.2   Notation and classifications for multiple membership structures

Using the general notation introduced for cross-classifications in Chapter 11 we can write (12.1) as

$$y_i = (X\beta)_i + \sum_{j \in school(i)} w_{i,j}^{(2)} u_j^{(2)} + e_i$$

$$u_j^{(2)} \sim N(0, \sigma_{u(2)}^2), \quad e_i \sim N(0, \sigma_e^2) \tag{12.3}$$

$$school(i) \in (1, \dots, J), \quad i = 1, \dots, N$$

where we have omitted the superscript (1) for the level 1 classification for convenience. The classification diagram for (12.3) is given in Figure 12.1.

The double arrow indicates a multiple membership relationship. For likelihood-based estimation we set up a set of dummy variables where the value is $w_{ij}^{(2)}$ for the j-th unit with a non-zero weight for level 1 unit i, and zero otherwise. As in the cross-classified case these then have random coefficients with variances constrained to be equal (see Appendix 11.1). For MCMC estimation the weights are incorporated into the

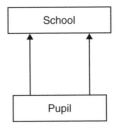

Figure 12.1   Classification diagram for 2-level multiple membership structure.

(multivariate) normal distribution used for sampling residuals at each iteration. Details are given in Browne et al. (2001).

We now illustrate the analysis of a multiple membership model with an example of veterinary data.

12.3 An example of salmonella infection

The data concern outbreaks of salmonella infection in Danish chicken flocks between 1995 and 1997. The response is whether or not salmonella is present in a (slaughtered) 'child' flock. Each flock is kept in a house within a farm, the hierarchical component, and is created from a mixture of up to six parent flocks, the multiple membership component. There is also the cross-classification of the parent flocks by house within farms. In addition there are four chicken hatcheries for which we fit three dummy variables, together with two dummy variables for year. There are 200 parent flocks, 304 farms, 725 houses and 10 127 child flocks of chickens. The classification diagram is given in Figure 12.2 and the model is

$$y_i \sim Bin(1, \pi_i)$$

$$\text{logit}(\pi_i) = (X\beta)_i + u^{(2)}_{House(i)} + u^{(3)}_{Farm(i)} + \sum_{j \in Flock(i)} w^{(4)}_{i,j} u^{(4)}_j \tag{12.4}$$

$$u^{(2)}_{House(i)} \sim N(0, \sigma^2_{u(2)}), \quad u^{(3)}_{Farm(i)} \sim N(0, \sigma^2_{u(3)}), \quad u^{(4)}_j \sim N(0, \sigma^2_{u(4)})$$

The results of fitting (12.4) using MCMC are given in Table 12.1. The year 1995 and hatchery 1 are taken as base categories. Default inverse gamma priors are used for variances and MH estimation is based upon 50 000 iterations with a burn in of 20 000.

It is clear that most of the variation is between farms and between parent flocks, with some large hatchery differences. Residuals can be estimated for all effects so that extreme farms and parent flocks can be identified. Further details can be found in Browne et al. (2001).

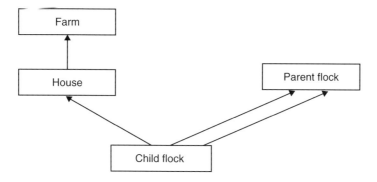

Figure 12.2 Classification diagram for salmonella data.

12.4   A repeated measures multiple membership model

Goldstein et al. (2000) consider repeated measures household data where individuals move from household to household over a period of 5 years divided into 6-monthly occasions. The response is the average length of time each individual has spent in all households since the start of the study, up to the current occasion. A household is not a constant unit but becomes a new one when any individual leaves or enters. The classification diagram is given in Figure 12.3 and the model in (12.5). The lowest level unit is the measurement occasion indexed by i, and these are nested within individuals. At each occasion the measurement has a contribution from a set of households so that occasions are formally multiple members of households.

$$y_i = (X\beta)_i + u^{(2)}_{individual(i)} + \sum_{j \in Household(i)} w^{(3)}_{i,j} u^{(3)}_j + e_i$$

$$u^{(2)}_{individual(i)} \sim N(0, \sigma^2_{u(2)}), \quad u^{(3)}_j \sim N(0, \sigma^2_{u(3)}), \quad e_i \sim N(0, \sigma^2_e)$$

(12.5)

The analysis given in Goldstein et al. shows a very large between-household variance compared to the between-individual variance, together with effects for age and nationality.

Table 12.1   MCMC estimates for salmonella data

	Estimate (s.e.)
Fixed	
Intercept	−2.33(0.22)
1996–1995	−1.24(0.17)
1997–1995	−1.16(0.19)
Hatchery 2–hatchery 1	−1.73(0.26)
Hatchery 3–hatchery 1	−0.20(0.25)
Hatchery 4–hatchery 1	−1.06(0.38)
Variances	
Parent flock	0.88 (0.18)
Farm	0.92 (0.20)
House	0.20 (0.11)

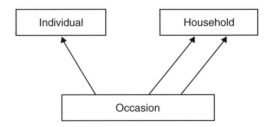

Figure 12.3   A repeated measures multiple membership model.

12.5   Individuals as higher level units

In all our examples of cross-classifications and multiple membership structures so far considered, individuals, or occasions in repeated measures models, are the lowest level units. In some structures, however, they can become higher level units with traditional higher level units becoming level 1 units. Consider the following educational structure.

Suppose we have students who are assessed individually and also in groups where the measurement is made at the group level and we wish to know what individuals contribute to the group response, and in particular the correlation between the individual assessment and this contribution. Suppose also that we have data where each individual in a sample is assessed, over time, in several groups with differing compositions. We have two responses and can write the following model

$$y_{1i}^{(1)} = (X_1\beta_1)_i + e_i^{(1)} + \sum_{j \in i} w_j u_{1j}^{(2)} \quad \text{(group level response)}$$

$$y_{2j}^{(2)} = (X_2\beta_2)_j + u_{2j}^{(2)} \quad \text{(individual level response)} \tag{12.6}$$

$$\text{cov}(u_{1j}^{(2)}, u_{2j}^{(2)}) \neq 0$$

The superscript indexes the classification, starting at level 1 which is now the group level so that groups are considered as multiple members of individuals: there are typically more groups than individuals. The classification diagram for (12.6) is given in Figure 12.4. The dotted line linking the two responses is used to indicate a multivariate model, with the vertical line separating the two components of the model.

This model allows us to estimate the individual level residuals $u_{1j}^{(2)}$, contributing to the group level response, and also the correlation between these and the individual level residuals associated with the individual level response $u_{2j}^{(2)}$. The model can be extended straightforwardly to incorporate random coefficients and discrete responses.

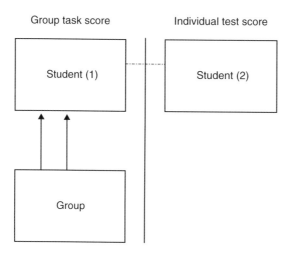

Figure 12.4   Classification diagram with two responses.

Other possible applications of such models are in applications where teams with varying compositions of individuals are formed in sporting activities or for work tasks.

12.6   Spatial models

In spatial modelling, measurements on individuals within an area are assumed to depend both on that area's effect and the effects of surrounding areas. Thus, suppose we are studying health status, we might start by assuming a 2-level model with an individual random effect and a random effect from the area in which a person lives. It may, however, be unrealistic to ignore the effects of surrounding areas (neighbours) on health status and to include these we might have a model such as the following

$$y_i = (X\beta)_i + \sum_{j \in neighbours(i)} w_{i,j}^{(2)} u_j^{(2)} + u_{area(i)}^{(3)} + e_i$$

(12.7)

$$u_j^{(2)} \sim N(0, \sigma_{u(2)}^2), \quad u_j^{(3)} \sim N(0, \sigma_{u(3)}^2), \quad e_i \sim N(0, \sigma_e^2)$$

We are here modelling separate random effects for the area of residence (classification 3) and the neighbouring areas (classification 2). In some cases all the areas in a sample are neighbours to all other areas so that for each area there are two random effects in a variance components model and we can fit a covariance between them. Langford et al. (1999) do this for some mortality data in Scotland and show a high correlation suggesting that areas have similar relative effects when they are contributing as neighbours and as the area of residence. In this particular case one might replace the two classifications by a single one with one random effect for each area. Leyland (2001) gives further examples and extends the model to consider a multivariate response where covariances between different responses are modelled for both area of residence and for neighbours. Browne et al. (2001b) compare the multiple membership spatial model with the traditional conditional autoregressive (CAR) model for an example with Poisson responses for disease prevalence data.

We can extend (12.7) for different response types and with further random coefficients. There are implicit weights in (12.7); equal weights attached to the neighbours, and also an equal weight for the total neighbourhood contribution and the residence area contribution. While the weights attached to neighbours are often chosen to be equal, other choices based upon distance are also possible. It is also possible to give more, or less, weight to the area of residence and both choices can be subjected to sensitivity analysis.

A further extension of these models is to consider as a further classification, say, the area where an individual works: if an individual works in more than one area then this will be another multiple membership classification.

These spatial models can be used to model interacting institutions. Thus, for example, the effects of schooling may come not only from the school that a student belongs to but also from neighbouring schools that may be competing for resources, etc. This would lead to a model such as (12.7) where each school will affect student outcomes for its own students and also for those in neighbouring schools. One of the aims of such a model might be to try to discover if there were characteristics of schools that could explain 'neighbouring' effects. In addition, of course, we can jointly model the effects of other classifications such as area of residence.

12.7 Missing identification models

A common problem in multilevel data is where the identification of a unit, usually at a higher level, is uncertain. Thus, for example we may have information from a household survey where, for some households, the area is unknown because the information has been lost or withheld. In a longitudinal study of schooling we may know that a student has changed school but not which school she came from. In such cases, if we can attach a *probability* of unit membership for each available higher level unit then we can use these probabilities in our model. If we denote these probabilities by $\pi_{i,j}$ then for a Normal variance components model we can write

$$y_i = (X\beta)_i + \sum_{j \in unit(i)} u_j^{(2)} \sqrt{\pi_{i,j}^{(2)}} + e_i$$

$$\text{var}\left(\sum_{j \in unit(i)} u_j^{(2)} \sqrt{\pi_{i,j}^{(2)}} \right) = \sigma_{u(2)}^2 \sum_j \left(\sqrt{\pi_{i,j}^{(2)}} \right)^2 = \sigma_{u(2)}^2$$

$$\text{cov}\left[\left(\sum_{j \in unit(i_1)} u_j^{(2)} \sqrt{\pi_{i_1,j}^{(2)}} \right) \left(\sum_{j \in unit(i_2)} u_j^{(2)} \sqrt{\pi_{i_2,j}^{(2)}} \right) \right] = \sigma_{u(2)}^2 \qquad (12.8)$$

$$\sum_{j \in unit(i_1) = j \in unit(i_2)} \sqrt{\pi_{i_1,j}^{(2)} \pi_{i_2,j}^{(2)}}$$

$$\sum_{j \in unit(i)} \pi_{i,j}^{(2)} = 1$$

Unlike the multiple membership model the level 2 variation is still $\sigma_{u(2)}^2$ since we are here assuming that students actually belong to just one school, so that the true level 2 variation is unchanged. Hill and Goldstein (1998) give further details and an example. In the common situation where there is complete ignorance about membership we can set $\pi_{i,j} = J$, the number of level 2 units. In this case the covariance between two such students becomes $\sigma_{u(2)}^2/J$ with corresponding correlation $\sigma_{u(2)}^2/J(\sigma_{u(2)}^2 + \sigma_e^2)$ which reflects the fact that the probability of two randomly chosen students belonging to the same school is $1/J$. Estimation for models such as (12.8) is similar to that for multiple membership models.

We can have models which are mixtures of multiple membership structures and missing identifications. For example, in a study of examination results we may know that a student has moved schools at a particular time but not know the identity of the previous school. In this case we would specify a set of identification probabilities for the previous school which would then be multiplied by the multiple membership weight for that period of schooling.

Another application of these models is where the identification information is missing, at least partly, by design. Thus, for example, the release of survey or census data is subject to confidentiality safeguards and in particular this often implies the removal of locality identifications. In such cases, data could be provided where, for each actual locality, units would be assigned to it from the complete set of all localities with known probabilities. Thus, attached to each lowest level unit in a locality would be a set of probabilities of belonging to each member of the full set of localities. In practice only a small number of such probabilities would be non-zero for each lowest level unit. If the probabilities are suitably chosen, this could satisfy confidentiality requirements since the actual locality would only be known probabilistically. The data could then be analysed using missing identification models since the probabilities are fully known for each unit.

13
Measurement Errors in Multilevel Models

13.1 A basic measurement error model

Many measurements are made with substantial error components, especially in the social and biological sciences. If the measurement were to be repeated we would not expect always to get an identical result. In some cases, such as the measurement of individual height or weight, the errors may be so small that they can safely be ignored in practice. In other cases, for example for educational tests and attitude measures, this usually will not be true and a failure to ignore errors may lead to incorrect inferences. Fuller (1987) provides a comprehensive account of methods for dealing with measurement errors in linear models and this chapter extends some of those procedures to the multilevel model and discusses some of the difficulties that arise when dealing with multilevel data.

A basic model for measurement errors in a 2-level continuous response model for the h-th explanatory variable and the response is as follows

$$Y_{ij} = y_{ij} + q_{ij}$$
$$X_{hij} = x_{hij} + m_{hij} \tag{13.1}$$
$$\text{cov}(q_{ij}q_{i'j}) = \text{cov}(m_{hij}m_{hi'j}) = 0$$

We adopt the convention that upper case letters denote the observed measurements and lower case the underlying 'true' measurements. Thus, we can think of these true measurements as being the expected values of repeated measurements of the same unit where the measurement errors are independent and are also independent of the true values. We define the *reliability* of the h-th explanatory variable

$$R_h = \sigma_{hx}^2/\sigma_{hX}^2 = (\sigma_{hX}^2 - \sigma_{hm}^2)/\sigma_{hX}^2 \tag{13.2}$$

that is the variance of the true values divided by the variance of the observed values.

This immediately raises two problems. When we are measuring such things as attitudes or educational achievement, we cannot carry out repeat measurements to obtain estimates of the σ_{hm}^2 because the measurement errors cannot be assumed to be independent. Another way of viewing this is to say that the process of measurement itself has changed the individual being measured, so that the underlying true value has also changed.

The second problem is that we have to define a suitable population. The definition of reliability is population dependent, so that, for example, if the measurement error variance σ_{hm}^2 remains constant but the population heterogeneity of the true values increases then the reliability will increase. Thus, the reliability may be lower within population subgroups, defined by social status say, than in the population as a whole. In particular, the reliability of a test score may be smaller within level 2 units, say schools, than across all students.

In this chapter we shall assume that the variances and covariances of the measurement errors are known, or rather that suitable estimates exist, and where possible, with estimates of their precision. At the very least sensitivity analysis should be carried out. The topic of measurement error estimation is a complex one, and there are in general no simple solutions, except where the assumption of independence of errors on repeated measuring can be made. The common procedure, especially in education, of using 'internal' measures based upon correlational patterns of test or scale items, is unsatisfactory for a number of reasons and may often result in reliability estimates that are too high. Ecob and Goldstein (1983) discuss these and propose some alternative estimation procedures. McDonald (1985) and other authors discuss the exploration and estimation of measurement error variances within a structural equation model, which has much in common with the suggestions of Ecob and Goldstein (1983).

13.2　Moment-based estimators

All the models in this section assume that the measurements that contain measurement errors do not have random coefficients. Details are given in Appendix 13.1. The random coefficient case will be dealt with when we discuss MCMC estimation.

13.2.1　Measurement errors in level 1 variables

We use a two level model to show how measurement errors can be incorporated into an analysis. We write for the true model

$$y_{ij} = (x\beta)_{ij} + (z_u u)_j + (z_e e)_{ij} \tag{13.3}$$

We assume that it is this true model for which we wish to make estimates. In some situations, for example where we wish simply to make a prediction for a response variable based upon observed values, then it is appropriate to treat these without correcting for measurement errors. If we wish to understand the nature of any underlying relationships, however, we require estimates for the parameters of the true model.

For the observed variables (13.3) gives

$$Y_{ij} = q_{ij} - (m\beta)_{ij} + (X\beta)_{ij} + (z_u u)_j + (z_e e)_{ij} \tag{13.4}$$

In Appendix 13.1 we show that the fixed effects are estimated by

$$\hat{\beta} = \hat{M}_{xx}^{-1} \hat{M}_{xy}$$
$$\hat{M}_{xx} = X^T \hat{V}^{-1} X - C_{\Omega_1} \tag{13.5}$$
$$C_{\Omega_1} = \left\{ \sum_i \sigma^{ii} \sigma_{(h_1, h_2)m}^i \right\}$$

where $\sigma^i_{(h_1,h_2)m}$ is the covariance between the measurement errors for explanatory variables h_1, h_2 for the i-th level 1 unit. The last expression in (13.5) is a correction matrix for the measurement errors and has elements which are weighted averages of the covariances of the measurement errors for each level over all the level 1 units in the sample with the weights being the diagonal elements of V^{-1}. In variance component models this is a simple average over the level 1 units, and in the common case where the covariance matrix of the measurement errors is assumed to be constant over level 1 units, we have

$$C_{\Omega_1} = tr(V^{-1})\Omega_{1m}, \quad \Omega_{1m} = \{\sigma_{(h_1,h_2)m}\} \tag{13.6}$$

An approximation to the covariance matrix of the estimates is given in Appendix 13.1, as is an expression for the estimation of the random parameters. For the constant measurement error covariance case with no measurement errors in the response variable this covariance matrix is given by

$$\hat{M}_{xx}^{-1}(X^T \hat{V}^{-1}X + X^T \hat{V}^{-2}T_{1m}X)\hat{M}_{xx}^{-1}$$
$$T_{1m} = (\hat{\beta}^T \Omega_{1m}\hat{\beta})I_{(n)} \tag{13.7}$$

and in the estimation of the random parameters the term T_{1m} is subtracted from $\tilde{Y}\tilde{Y}^T$ at each iteration. It is important in some applications to allow the measurement error variances to vary as a function of explanatory variables. For example, in perinatal studies, the measurement of gestation length may be quite accurate for some pregnancies where careful records are kept but less so in others.

13.2.2 Measurement errors in higher level variables

Where variables are defined at level 2 or above with measurement errors we have analogous results, with details given in Appendix 13.1. Thus the correction term to be used in addition to C_{Ω_1} with a constant measurement error covariance matrix in a 2-level model is

$$C_{\Omega_2} = \left(\sum_j J_{n_j}^T V_j^{-1} J_{n_j}\right)\Omega_{2m} \tag{13.8}$$

where J_n is a vector of ones of length n and V_j is the j-th block of V.

A case of particular interest is where the level 2 variable is an aggregation of a level 1 variable. Woodhouse et al. (1996) consider this case in detail and give detailed derivations. Consider the case where we have a level 2 variable which is the mean of a level 1 variable

$$X_{1\cdot j} = \frac{1}{n_j}\sum_i X_{1ij}$$

The variance over the whole sample is therefore given by

$$\text{var}(X_{1 \cdot j}) = n_j \frac{\text{var}(X_{1ij})}{n_j^2} + n_j(1 - n_j)\frac{\text{cov}(X_{1ij}X_{1i'j})}{n_j^2}$$

$$= \frac{1}{n_j}\sigma_{(1)}^2(X_1) + \frac{n_j - 1}{n_j}\sigma_{(2)}^2(X_1) \qquad (13.9)$$

where we assume constant variances and covariances within level 2 units for the X_{1ij}. The number of level 1 units actually measured in the j-th level 2 unit is n_j out of a total number of units N_j. Straightforward estimates of the parameters can be obtained by carrying out a variance components analysis with X_{1ij} as response, fitting only the overall mean in the fixed part, so that the covariance is the level 2 variance estimate.

For the true values we have an analogous result where now we consider the variance of the mean of the true values for *all* the level 1 units in each level 2 unit. There are, in effect, two sources of error in $X_{1 \cdot j}$. There is the error inherent in the level 1 measurement X_1 which is averaged across the level 1 units in each level 2 unit and there is the sampling error which occurs when $n_j < N_j$, that is not all the units in the level 2 unit are measured. Thus the true value is the average for all the level 1 units in each level 2 unit of the true level 1 measurements. Since the measurement errors are assumed independent we have

$$\text{var}(x_{1 \cdot j}) = \frac{1}{N_j}\sigma_{(1)}^2(x_1) + \frac{N_j - 1}{N_j}\sigma_{(2)}^2(x_1) \qquad (13.10)$$

This gives us the following expression for the required measurement error variance for the aggregated variable

$$\sigma_{1 \cdot m}^2 = \left(\frac{1}{n_j} - \frac{R_1}{N_j}\right)\sigma_{(1)}^2(X_1) - \left(\frac{1}{n_j} - \frac{1}{N_j}\right)\sigma_{(2)}^2(X_1) \qquad (13.11)$$

where the reliability R_1 is estimated from the level 1 variation.

If both the level 1 observed variable and its aggregate are included as explanatory variables then clearly their measurement errors are correlated and the correlation is given by

$$\frac{1 - R_1}{n_j}\sigma_{(1)}^2(X_1)$$

In the expressions for the correction matrices, we have considered the separate contributions from levels 1 and 2. Where there is a 'cross-level' correlation between measurement errors as above then we add the level 1 variable to Ω_{2m} using (13.11) for the covariance together with a zero variance. The measurement error variance for the level 1 explanatory variable becomes a component of Ω_{1m}. A detailed derivation of these results is given by Woodhouse et al. (1996).

13.3 A 2-level example with measurement error at both levels

We use the Junior School Project data reading score at the age of 11 years as our response with the 8-year mathematics score as predictor, fitting also social class (Non-manual and Manual) and gender. There are 728 students in 48 schools in this analysis. The

Table 13.1 Eleven-year Normalized mathematics score related to Normalized 8-year score, gender and social class for different 8-year score level 1 reliabilities; adjusting for measurement errors in the 8-year score

Parameter	Estimate (s.e.)		
	A ($R_1 = 1.0$)	B ($R_1 = 0.9$)	C ($R_1 = 0.8$)
Fixed			
Intercept	0.13	0.10	0.07
8-year score	0.670 (0.026)	0.750 (0.030)	0.860 (0.0035)
Gender	−0.048 (0.050)	−0.048 (0.050)	−0.047 (0.052)
Non-manual	0.140 (0.060)	0.100 (0.060)	0.060 (0.060)
Random			
σ_u^2	0.080 (0.023)	0.081 (0.024)	0.081 (0.024)
σ_e^2	0.422 (0.023)	0.373 (0.023)	0.310 (0.025)
Variance partition coefficient	0.16	0.18	0.21

scores at age 11 and 8 have been transformed to have a standard Normal distribution as in Table 2.7. We shall allow for measurement errors in both the test scores.

In the original analyses of these data (Mortimore et al, 1988) reliabilities are not given, and for the reasons given above are unlikely to be well estimated. For the purpose of our analyses we investigate a range of reliabilities from 0.8 to 1.0 to study the effect of introducing increasing amounts of measurement error.

It can be seen in Table 13.1 that the inferences about the fixed parameters and the level 1 variance and variance partition coefficient change markedly in moving from an assumption of zero measurement error to a reliability of 0.8. The increase in the variance partition coefficient reflects the fact that it is only the level 1 variance which decreases as the reliability decreases. The difference between the children from non-manual and manual backgrounds is reduced considerably as the reliability decreases.

We now look at the effect of adjusting additionally for measurement error in the response variable. To illustrate this we look at the effects on the individual parameters for a range of values for the reliabilities of both response and explanatory variables.

As shown in Table 13.2, as the response variable reliability decreases, so does the level 1 variance estimate. Likewise, as the reliability of the 8-year score decreases the level 1 variance decreases. The combined effect of both reliabilities being 0.8 produces a variance that is a quarter of the estimate that assumes no unreliability. When both the reliabilities reach the value of 0.7 the level 1 variance decreases to zero! By contrast, the level 2 variance is hardly altered. For the coefficient of the 8-year score and social class the greatest change is with the reliability of the 8-year score.

As the reliability decreases so the strength of the relationship with 8-year score increases, while the social class difference decreases substantially. The gender difference is changed very little.

Clearly, the requirement of a positive level 1 variance implies particular lower bounds on the reliabilities and measurement error variances, and underlines the importance of obtaining good estimates of these parameters or at least a range of reasonable estimates. The range of variance partition coefficient values, from 16% to 21% also indicates that we need to take care in interpreting small values of such coefficients without adjusting for measurement error. This has important implications for 'school effectiveness' studies (Goldstein, 1997) where ignoring measurement errors can be expected to lead to underestimation of school effects.

Table 13.2 Parameter estimates (standard errors) for values of explanatory and response variables

Eight year score reliability	Response reliability		
	1.0	0.9	0.8
Eight-year score			
1.0	0.67	0.67	0.67
0.9	0.75	0.75	0.75
0.8	0.86	0.86	0.87
Gender			
1.0	−0.048	−0.048	−0.047
0.9	−0.048	−0.048	−0.047
0.8	−0.047	−0.046	−0.046
Non-manual			
1.0	0.14	0.14	0.15
0.9	0.10	0.10	0.11
0.8	0.06	0.06	0.06
Level 2 variance			
1.0	0.080	0.079	0.078
0.9	0.081	0.080	0.079
0.8	0.081	0.080	0.079
Level 1 variance			
1.0	0.422	0.324	0.225
0.9	0.373	0.274	0.176
0.8	0.310	0.211	0.112

13.4   Multivariate responses

To model multivariate data, we specify a dummy (0,1) variable for each response and corresponding interactions with other explanatory variables. Then $C_{\Omega_1}, C_{\Omega_2}$ in (13.5) and (13.8) are modified so that for each level 1 or level 2 unit, the covariance between measurement errors is set to zero when either of the corresponding dummy variables is zero and likewise for the variances. This is equivalent to specifying the same covariance matrix of measurement errors for each set of explanatory variables corresponding to a response variable, with no covariances across these sets. For the response variables we likewise specify the separate measurement error variances for each one using the general procedures in Appendix 13.1.

13.5   Nonlinear models

Consider the 2-level model (8.3) in Chapter 8 where there are measurement errors in the explanatory variables for the fixed part of the model. In this case we can obtain an approximate analysis by using the observed values in the updating formulae and replacing the measurement error covariances in (13.5) by

$$(f'_{(i)})^2 \sigma^i_{(h_1, h_2)m} \tag{13.12}$$

where $f'_{(i)}$ is the first differential of the nonlinear function for the i-th level 1 unit with a corresponding expression for level 2 measurement errors. The derivation of (13.12) is given in Appendix 13.1.

13.6   Measurement errors for discrete explanatory variables

Assume that we have a categorical explanatory variable with r categories. We shall consider only a single such variable, since multiple variables can in principle be handled by considering the p-way table based upon them as a single vector. In practice it will often be reasonable to assume that their measurement errors are uncorrelated so that they can be considered separately. Likewise we can often assume that measurement errors in discrete explanatory variables are uncorrelated with those in continuous variables. The following derivations parallel those given by Fuller (1987, section 3.4). We consider only level 1 explanatory variables, but the extension to higher levels follows straightforwardly.

Let $A_{(i)}(1 \times r)$ be a row vector for the i-th level 1 unit containing a one for the category which is observed and zeros elsewhere. Let k_{mn} be the probability that a level 1 unit with true category n is observed in category m. We write

$$K = \{k_{mn}\}, \quad \text{where } K_m \text{ is the } m\text{-}th \text{ column of } K$$

and define

$$X_{(i)}^T = K^{-1}A_{(i)}^T \tag{13.13}$$

If $x_{(i)}$ is the true value we write

$$A_{(i)} = x_{(i)} + \varepsilon_{(i)}, \quad E(A_{(i)}|x_{(i)}) = x_{(i)}K^T$$

We also write

$$X_{(i)} - x_{(i)} = m_{(i)}$$

so that

$$E(m_{(i)}|x_{(i)}) = 0$$

which gives the familiar form for the errors in variables model where the unknown true value $x_{(i)}$ is uncorrelated with the measurement error. The $X_{(i)}$ become the new set of observed values and interest is in the regression on the true category values $x_{(i)}$. The vector $x_{(i)}$ consists of a single value of one and the remainder zero. We have

$$\mathrm{cov}(A_{(i)}^T|x_{(i)} = l_m) = \Sigma_{(i)(m)} = diag(K_m) - K_m K_m^T$$

where l_m is an r-dimensional vector with 1 in the m-th position and zeros elsewhere. For the i-th level 1 unit define

$$\Omega_{(i)m} = \mathrm{cov}(m_{(i)}^T|x_{(i)} = l_m) = K^{-1}\Sigma_{(i)(m)}K^{-1^T} \tag{13.14}$$

and we use as our estimate of the covariance matrix of measurement errors the matrix in (13.14) conditional on the observed $A_{(i)}$.

$$\hat{\Omega}_{(i)m} = \Omega_{(i)m}\left[\frac{P(x_{(i)} = l_m)}{P(A_{(i)} = l_m)}\right] \tag{13.15}$$

Table 13.3 Eleven-year normalized mathematics score related to Normalized 8-year score, gender and social class for different 8-year score level 1 reliabilities; adjusting for measurement errors in the 8-year score. MCMC estimates

Parameter	Estimate (s.e.)		
	$A\,(R_1 = 1.0)$	$B\,(R_1 = 0.9)$	$C\,(R_1 = 0.8)$
Fixed			
Intercept	0.14	0.13	0.13
8-year score	0.67 (0.027)	0.76 (0.04)	0.90 (0.04)
Gender	−0.039 (0.050)	−0.039 (0.050)	−0.041 (0.051)
Non-manual	0.15 (0.060)	0.15 (0.060)	0.15 (0.060)
Random			
σ_{u0}^2	0.093 (0.023)	0.091 (0.026)	0.091 (0.026)
σ_{u01}	−0.020 (0.010)	−0.018 (0.010)	−0.016 (0.011)
σ_{u1}^2	0.014 (0.008)	0.012 (0.007)	0.013 (0.007)
σ_e^2	0.413 (0.023)	0.362 (0.023)	0.270 (0.026)

The term in square brackets can be estimated as follows. If μ_A, μ_x are the observed and true vectors of probabilities for the categories, then

$$\mu_x = K^{-1}\mu_A$$

and given the sample estimate of μ_A we can estimate μ_x. The estimate given by (13.15) is then used as in the case of continuous explanatory variables measured with error. In the general model the number of explanatory variables will generally be one less than the number of categories, with one of the categories chosen as the base and omitted.

In practice, the matrix of probabilities K, is normally assumed constant but can itself depend on further explanatory variables. Often we will not have a good estimate of it, and we may need to make some simplifying assumptions. In the case of a binary variable it may be possible to assume equal misclassification probabilities k, in which case only a single value needs to be determined, and in practice a range of values can be explored.

13.7   MCMC estimation for measurement error models

Appendix 13.1 sets out the steps involved in modelling measurement error data using MCMC. It involves adding extra steps to the existing Gibbs or MH algorithms and assuming particular distributions for the true scores and measurement errors. We now apply this to the data in Table 13.1, including also a random coefficient model.

As Table 13.3 shows, the results are similar to those obtained for the variance components model when increasing amounts of measurement error are introduced, with a somewhat greater effect on the level 1 variance estimate. The major advantage of MCMC estimation in this case is the ability to handle measurement errors in explanatory variables with random coefficients.

Where discrete predictor variables are involved we need to consider the probability that a value observed in a particular category actually comes from each possible category; that is we need to know misclassification probabilities and further work along these lines is being carried out.

Appendix 13.1

Measurement error estimation

13.1.1 Moment-based estimators for a basic 2-level model

We consider a 2-level model and write for the response and explanatory variables

$$Y_{ij} = y_{ij} + q_{ij}$$
$$X_{hij} = x_{hij} + m_{hij}$$
$$\text{cov}(q_{ij}q_{i'j}) = \text{cov}(m_{hij}m_{hi'j}) = 0 \qquad (13.1.1)$$
$$E(q_{ij}) = E(m_{hij}) = 0$$
$$\text{cov}(m_{h_1ij}m_{h_2ij}) = \sigma^i_{(h_1h_2)jm}$$

for the h-th explanatory variable with measurement error vector m_h and with q as the measurement error vector for the response. We use upper case for the observed and lower case for the 'true' values which are the expected values of the observed measurements. Each level 1 unit may have its own set of measurement error variances. Where we have a level 2 explanatory variable, then the measurement error is constant within a level 2 unit.

We write the 'true' model in the general form

$$y_{ij} = (x\beta)_{ij} + (z_u u)_j + (z_e e)_{ij} \qquad (13.1.2)$$

which gives the model for the observed variables as

$$Y_{ij} = q_{ij} - (m\beta)_{ij} + (X\beta)_{ij} + (z_u u)_j + (z_e e)_{ij} \quad m = \{m_h\} \qquad (13.1.3)$$

For the true values write

$$M_{xx} = x^T V^{-1} x, \quad M_{xy} = x^T V^{-1} y \quad \hat{\beta} = M_{xx}^{-1} M_{yy} \qquad (13.1.4)$$

Now

$$X^T V^{-1} X = (x + m)^T V^{-1} (x + m)$$
$$= x^T V^{-1} x + m^T V^{-1} x + x^T V^{-1} m + m^T V^{-1} m \qquad (13.1.5)$$

so that

$$E(X^T V^{-1} X) = x^T V^{-1} x + E(m^T V^{-1} m) \qquad (13.1.6)$$

If we further assume that q and m are uncorrelated then we have

$$E(X^T V^{-1} Y) = x^T V^{-1} y \qquad (13.1.7)$$

Thus, to estimate the fixed parameters we require $E(m^T V^{-1} m)$ and we now consider how to obtain this for measurement errors at both level 1 and level 2.

13.1.2 Parameter estimation

For errors of measurement in level 1 units the (h_1, h_2) element of $E(m^T V^{-1} m)$ is

$$\sum_{i=1}^{N} \sigma^{ii} \sigma^{i}_{(h_1 h_2) jm}$$

$$\text{with } C_{\Omega_1} = \left\{ \sum_i \sigma^{ii} \sigma^{i}_{(h_1 h_2) jm} \right\} \qquad (13.1.8)$$

where N is the total number of level 1 units. In the case where each level 1 unit has the same covariance matrix of measurement errors we have

$$C_{\Omega_1} = tr(V^{-1}) \Omega_{1m}, \quad \Omega_{1m} = \left\{ \sigma_{(h_1 h_2) m} \right\} \qquad (13.1.9)$$

For errors of measurement in level 2 explanatory variables we have

$$C_{\Omega_2} = \sum_j (J_{(1,n_j)} V_j^{-1} J_{(n_j,1)}) \Omega_{2jm} \qquad (13.1.10)$$

Where Ω_{2jm} is the covariance matrix of measurement errors for the j-th level 2 block, and $J_{(r,s)}$ is a $(r \times s)$ matrix of ones. In Chapter 13 we discuss how to obtain the Ω_{2jm} for level 2 variables which are aggregates of level 1 variables.

For the measurement error corrected estimate of the fixed coefficients we have

$$\hat{M}_{xx} = M_{XX} - C_{\Omega_1} - C_{\Omega_2} \qquad (13.1.11)$$

For the random component based upon the model with observed variables write the residual $v_{ij} = (z_u u)_j + (z_e e)_{ij} + q_{ij} - (m\beta)_{ij}$, $v = \{v_{ij}\}$ which gives

$$E(vv^T) = V + \bigoplus_{ij} \sigma^2_{ijq} + T_1 + T_2$$

$$T_1 = \bigoplus_{ij} (\hat{\beta}^T \Omega_{1ijm} \hat{\beta}), \quad T_2 = \bigoplus_j (\hat{\beta}^T \Omega_{2jm} \hat{\beta}) J_{(n_j,n_j)} \qquad (13.1.12)$$

where σ^2_{ijq} is the measurement error variance for the ij-th response measurement. Thus the quantity $\bigoplus_{ij} \sigma^2_{ijq} + T_1 + T_2$ should be subtracted from the sum of products matrix $\tilde{Y}\tilde{Y}^T$ at each iteration, when estimating the random parameters.

The covariance matrix of the estimated fixed coefficients is given by

$$\hat{M}_{xx}^{-1}(X^T V^{-1} X + X^T V^{-1} Q V^{-1} X + X^T V^{-2}[T_1 + T_2]X)\hat{M}_{xx}^{-1}$$

$$Q = \bigoplus_{ij} \sigma^2_{ijq} \qquad (13.1.13)$$

This expression ignores any variation in the estimation of the measurement error variance itself, although Goldstein (1986) includes terms for this.

13.1.3 Random coefficients for explanatory variables measured with error

The above expressions assume that the coefficients of variables with measurement error are not random. Where such coefficients are random the above formulae do not apply, and in particular $m^T V^{-1} m$ has measurement errors in all its components. Woodhouse (1998) discusses this problem in detail and suggests that moment-based estimators are not really feasible.

13.1.4 Nonlinear models

Consider first the case where just the fixed part explanatory variables have measurement errors at level 1 in the single component 2-level nonlinear model for the i-th level 1 unit

$$y_{(i)} = f_{(i)}(X\beta + random)$$

which yields the linearization

$$y_{(i)} - \left\{ f_{(i)}(X\beta) - \sum_k \beta_{k,t} x^*_{(i)k} \right\} = \sum_k \beta_{k,t+1} x^*_{(i)k} + random\ terms \qquad (13.1.14)$$

where the explanatory variables are the observed measurements and the coefficients are the required ones corrected for measurement error and $x^*_{(i)k} = f'_{(i)} x_{(i)k}$. Consider the expansion of $f_{(i)}$ for the measurement error terms, to a first order approximation,

$$f_{(i)} = f_{(i),m_k=0} + \sum_k f'_{(i),u_k=0} m_{(i)k} \beta_{k,t} \qquad (13.1.15)$$

Thus we can use the observed explanatory variables with measurement error as an approximation to the use of the true values in the updating formulae, with $(f'_{(i)})^2 \sigma^i_{(h_1,h_2)m}$ replacing $\sigma^i_{(h_1,h_2)m}$ in (13.5).

13.1.5 MCMC estimation for measurement error models

We consider the case of a 2-level model with measurement errors in the explanatory variables, using Gibbs sampling. The model now allows for variables with measurement errors to have random coefficients. Using the same notation as before we write

$$y_{ij} = \beta_0 + (\beta_1 + u_{1j})x_{1ij} + u_{0j} + e_j$$
$$X_{1ij} = x_{1ij} + m_{1ij} \qquad (13.1.16)$$
$$m_{1ij} \sim N(0, \sigma_m^2), \quad x_{1ij} \sim (\theta, \phi^2)$$

ϕ^2 as a general scaled inverse chi squared prior with parameters υ_ϕ and S_ϕ^2. Note that we have introduced a distribution for the true values. This is analogous to the assumptions we made for the moment-based estimators where we derived the reliability as the ratio of the variance for the measurement errors and the sum of the variances from the distribution of the true values plus the measurement errors. Where there are further explanatory variables we simply add further steps for each of these in the algorithm, noting that for some there may be zero measurement errors. If the measurement errors are correlated we will need to modify the algorithm by defining a multivariate Normal distribution for the measurement errors.

We now set up a Gibbs sampler where the first four steps are as given in Appendix 2.4 and we add the following steps

- **Step 5** Sample new true values x from

$$p(x|Y \text{ and current parameters}) \sim N(\hat{x}_{1ij}, \hat{V}_{1ij})$$

$$\hat{V}_{1ij} = [(\beta_1 + u_{1j})^2 \sigma_e^{-2} + \sigma_m^{-2} + \phi^{-2}]^{-1} \tag{13.1.17}$$

$$\hat{x}_{1ij} = \hat{V}_{1ij}[(\beta_1 + u_{1j})(y_{ij} - \beta_0 - u_{0j})\sigma_e^{-2} + X_{1ij}\sigma_m^{-2} + \theta\phi^{-2}]$$

- **Step 6** Sample a new mean for the true values from

$$p(\theta|x_{1ij}, \phi^2) \sim N(\hat{\theta}, \hat{V}_\theta)$$

$$\hat{V}_\theta = \phi^2/N \tag{13.1.18}$$

$$\hat{\theta} = \left(\sum_{ij} x_{1ij}\right) \bigg/ N$$

N is number of level 1 units

- **Step 7** Sample a new variance parameter for the true values from

$$p(\phi^{-2}|x_{1ij}, \theta) \sim gamma(a_\phi, b_\phi)$$

$$a_\phi = (N + v_\phi)/2, \quad b_\phi = (v_\phi s_\phi^2 + \sum_{ij} (x_{1ij} - \theta)^2)/2 \tag{13.1.19}$$

- **Step 8** Obtain level 1 residuals by subtraction.

For errors of measurement in the response variable an extra step is required to sample true values of the response. Browne et al. (2001a) give further details and some simulation results.

In the above we have assumed default priors with a uniform for the mean of the true values and an inverse Gamma for the variance. We have also assumed that σ_m^2 is known, whereas in practice it will almost always be estimated. We could therefore place a prior distribution on this with parameters obtained, for example, from previous research.

For more complex models such as cross-classifications and multiple membership models and for models with discrete responses the above steps can be added to the existing algorithms.

For discrete explanatory variables the MCMC algorithm would need to consider misclassification probabilities and to use the probabilities of a value being in each category having observed a particular category.

14
Missing Data in Multilevel Models

14.1 A multivariate model for handling missing data

A characteristic of many studies is that some of the intended measurements are unavailable. In surveys, for example, this may occur through chance or because certain questions are unanswered by particular groups of respondents. We are concerned principally with missing values in the explanatory or predictor variables in a multilevel model, although the procedures can incorporate missing response variable values as is shown in the next section. An important distinction is made between situations where the existence of a missing data item can be considered a random event and where it is informative and the result of a non-random mechanism. Randomly missing data may be missing 'completely at random' or 'at random' conditionally on the values of other measurements. We shall begin with these two types of random event. Where data cannot be assumed to be missing at random, one approach is to attempt to model the missingness mechanism along with the model of interest, and this will be considered in a later section.

We consider the problem in two parts. First we develop a procedure for predicting data values which are missing and then we study ways to obtain model parameter estimates from the resulting 'filled-in' or 'completed' dataset. The prediction will use those measurements which are available, so that data values which are missing at random conditional on these measurements can be incorporated. Detailed discussions of missing data procedures are given by Rubin (1987) and Little (1992).

The basic exposition will be in terms of a single-level model for simplicity, pointing out the extensions for multilevel models.

14.2 Creating a completed data set

Consider the ordinary linear model

$$y_i = \beta_0 + \beta_1 x_{1i} + \beta_2 x_{2i} + e_i \tag{14.1}$$

for the i-th unit in a single-level model. Suppose that some of the x_{1i} are missing completely at random (MCAR) or conditionally missing at random (MAR) conditional on x_2. Label these unknown values x_{1i}^*. We consider the estimation of these by predicting them from the remaining observations and the parameter set θ for the prediction model, namely

$$\hat{x}_{1i}^* = E(x_{1i}^*|x_{2i}, y_i, \theta) \tag{14.2}$$

Where we have multivariate Normal data the prediction (14.2) is simply the linear regression of X_1 on X_2, Y, where the coefficients of this regression prediction are obtained from efficient, for example maximum likelihood, estimates of the parameters of the multivariate Normal distribution. This can be achieved using the procedures for modelling multivariate data described in Chapter 6 where the original level 1 units become new level 2 units and the multiple responses form the new level 1 units. We shall consider the case of non-Normal data later.

We define a multivariate model with three response variables, Y, X_1, X_2 and three corresponding dummy variables, say Z_0, Z_1, Z_2 as described in Chapter 6. Some level 2 units will have all three response variables, but others will have only two where X_1 is missing. Write this as the 2-level model

$$v_{ij} = \beta_{0j}z_{0ij} + \beta_{1j}z_{1ij} + \beta_{2j}z_{2ij},$$

$$\tag{14.3}$$

$$\beta_j \sim MVN(\mu, \Omega), \quad \mu^T = (\mu_Y, \mu_1, \mu_2), \quad \Omega = \begin{pmatrix} \sigma_Y^2 & & \\ \sigma_{YX_1} & \sigma_{X_1}^2 & \\ \sigma_{YX_2} & \sigma_{X_1X_2} & \sigma_{X_2}^2 \end{pmatrix}$$

This model will produce efficient (ML in the Normal case) estimates of the coefficients in (14.1), namely

$$\begin{pmatrix} \beta_1 \\ \beta_2 \end{pmatrix} = \Omega_{XX}^{-1}\Omega_{XY}, \quad \Omega_{XX} = \begin{pmatrix} \sigma_{X_1}^2 & \\ \sigma_{X_1X_2} & \sigma_{X_2}^2 \end{pmatrix}, \quad \Omega_{XY} = \begin{pmatrix} \sigma_{YX_1} \\ \sigma_{YX_2} \end{pmatrix} \tag{14.4}$$

Also, for any missing value we can use the parameters from (14.4) to predict X_1 from X_2, Y. These predicted values are just the estimated new level 2 residuals from (14.3) for the missing values. Clearly this procedure extends to any number of variables with any pattern of missing data. We simply formulate the model as a multivariate response by introducing dummy variables for each variable and then estimating the residuals for the resulting 2-level model and choosing the appropriate residuals to fill in the missing values. This procedure extends in a straightforward way to multilevel data.

Suppose now we have a 2-level dataset with some explanatory variables measured at the original level 1 and some at the original level 2 and various values missing. We specify a 3-level multivariate response model where some of the responses are at level 2 and some at level 3. At level 2 of this model we estimate a covariance matrix for the original level 1 variables and at level 3 we estimate a covariance matrix for all the variables. For the original level 2 variables with missing values we estimate the residuals at level 3 and use these to fill in missing values. For the original level 1 variables we add the level 3 and the level 2 residuals together to obtain filled in values.

If we were to use the completed datasets in the usual way to fit a multilevel model the resulting estimates would be biased because the filled-in data are shrunken and have less variation than the original measurements. Little (1992) discusses this problem and in the next section we outline procedures for dealing with it.

14.3   Multiple imputation and error corrections

The usual multiple imputation (Rubin, 1987) procedure works as follows. The predicted values are adjusted to have their correct, on average, distributional properties by sampling from the multivariate distribution associated with the predicted values. Where we have, as in the above example, just one variable with missing values in a single level Normal model this involves a series of random draws from the Normal distribution with mean the residual estimate $\hat{x}_{1i}^*$ and variance given by the estimated (comparative) variance of this residual estimate. For small samples, in estimating this variance, we should also take account of the sampling variation of the estimated parameters, for example using a bootstrap procedure (Chapter 3) or the approximation in Appendix 2.2.

Where the residuals from two different levels are combined, as described above, several original level 1 units within the same original level 2 unit share the same level 2 residual so that we will need to sample from the multivariate distribution where the variances are simply the sums of the variances from the two levels and the common covariance is the variance of the original level 2 estimate. Where there are several variables with filled-in values then we need to sample from an extended multivariate distribution.

Having generated these 'corrections' we then fit our multilevel model in the usual way and obtain parameter estimates. This process is repeated m times, and the final estimates are suitably chosen averages of these sets of estimates.

For a parameter θ with point estimates θ_i and variance estimates $\text{var}(\theta_i)$, we form

$$\hat{\theta} = \sum_{i=1}^{m} \theta_i / m, \quad \text{var}(\hat{\theta}) = \bar{V} + (1 + m^{-1})B$$

$$\bar{V} = \sum_{i=1}^{m} \text{var}(\theta_i)/m, \quad B = (\theta_i - \hat{\theta})^2/(m-1)$$

(14.5)

with corresponding expressions for a set of parameters and their covariance matrix. The value of m is often chosen to be as small as 5. These final estimates are asymptotically efficient with consistent standard errors.

This kind of multiple imputation, in practice, has certain drawbacks. The principal one is the amount of computation required to carry out the several stages of analysis. As an alternative, the following procedure can be used.

For our simple example the imputation procedure implicitly assumes a model of the form

$$x_{1i} = \hat{x}_{1i}^* + w_{1i}$$ (14.6)

where the w_{1i} have the variances and covariances for the residuals estimated as above, and zero means. If we assume that the two terms on the right-hand side of (14.5) are uncorrelated, then we have

$$\text{var}(x_{1i}) = \text{var}(\hat{x}_{1i}^*) + \text{var}(w_{1i})$$

We see therefore that to obtain estimates for the fixed coefficients based upon the true values we can apply similar procedures to those used in the measurement error case (see Chapter 13) but with measurement error variances *added* rather than subtracted from the relevant quantities. Thus, using the notation in Appendix 13.1, for a 2-level model we have the following for a model with p explanatory variables with missing data at level 1. We form

$$\hat{M}_{xx} = X^{*^T} V^{-1} X^* + C_{\Omega_1} + C_{\Omega_2}$$

$$C_{\Omega_1} = \left\{ \sum_i \sigma^{ij} \sigma^{ij}_{e(h_1,h_2)w} \right\}$$ (14.7)

$$C_{\Omega_2} = \left\{ \sum_j (J^{*^T}_{n_j(h_1 h_2)} V_j^{-1} J^*_{n_j(h_1 h_2)}) \sigma^j_{uj(h_1 h_2)} \right\}$$

substituting sample estimates for the elements

$$C_{\Omega_1}(h_1, h_2) = \sum_i \sigma^{ij} \sigma^{ij}_{e(h_1,h_2)w}$$

$$C_{\Omega_2}(h_1, h_2) = \sum_j (J^{*^T}_{n_j(h_1 h_2)} V_j^{-1} J^*_{n_j(h_1 h_2)}) \sigma^j_{uj(h_1 h_2)}$$

For the ij-th level 1 unit σ^{ij} is the diagonal term of V^{-1} and $\sigma^{ij}_{e(h_1,h_2)w}$ is the corresponding covariance (or variance) between the (level 1) residuals for variables h_1, h_2 where these are both missing. The vector $J^*_{n_j(h_1 h_2)}$ contains a one if, for the j-th second level unit, variables h_1, h_2 are both missing and zero otherwise. The term $\sigma^j_{uj(h_1 h_2)}$ is the estimated covariance (or variance) between the (level 2) residuals for variables h_1, h_2.

The estimates of the fixed coefficients are given by

$$\hat{\beta} = \hat{M}_{xx}^{-1} \hat{M}_{xy}$$

The extensions for level 2 explanatory variables and discrete variables (see below) are likewise analogous to those described in Chapter 13.

In the single-level case for a single explanatory variable with missing data, these results reduce to the following. Order the completed data so that the imputed observations are grouped together first. Then, ignoring any correction for sampling variation, the adjustment is obtained by replacing $(X^T X)$ by

$$(X^T X) + \begin{pmatrix} n_1 \hat{\sigma}_w^2 & 0 \\ 0 & 0 \end{pmatrix}$$

where there are n_1 imputed values. This is very similar to the correction described by Beale and Little (1975), although these authors use an estimate based upon the observed residuals calculated from the complete data cases and approximate the covariance matrix by $\hat{M}_{xx}^{-1}$.

14.4   Discrete variables with missing data

Suppose we have one or more categorical explanatory variables as well as continuous variables with missing values. The first stage procedure is to obtain the predicted values. We can do this by treating all the variables together as a multivariate model

Table 14.1 JSP Mathematics data. Model A is full data analysis, model B omits cases with missing data, model C uses completed data, model D uses full missing data procedure of (14.7)

Parameter	Estimate (s.e.)			
	A	B	C	D
Fixed				
Constant	0.14	0.12	0.097	0.12
8-year score	0.095 (0.0037)	0.100 (0.0040)	0.105 (0.0037)	0.097 (0.0039)
Gender (boys–girls)	−0.044 (0.050)	−0.087 (0.054)	−0.067 (0.047)	−0.066 (0.051)
Social class (non-manual–manual)	0.154 (0.057)	0.113 (0.060)	0.107 (0.054)	0.135 (0.058)
Random				
Level 2: σ_{u0}^2	0.081 (0.023)	0.083 (0.025)	0.077 (0.022)	0.077 (0.023)
Level 1: σ_{e0}^2	0.423 (0.023)	0.415 (0.024)	0.378 (0.021)	0.412 (0.023)

with mixed continuous and discrete responses as described in Chapter 4. For each categorical variable we obtain the predicted probabilities of belonging to each category, corresponding to each dummy variable used in the subsequent analysis. For a single-level model these would be substituted to form the completed dataset. For a 2-level model we would add the level 3 residual from the initial multivariate model to each prediction. Thus, where the categorical variable is at level 1 then for each level 1 unit where variables are missing the dummy variable values are replaced by estimates. We can obtain the $\sigma_{e(h_1,h_2)w}^{ij}$ together with covariances between discrete and continuous variables from the model estimates (see Chapter 4) and the relevant higher level variances and covariances are added for models with further levels. Care is needed with such linear predictions for discrete data and further research is required to determine its usefulness.

14.5 An example with missing data

We use the Junior School Project dataset and model A of Table 13.1 to illustrate the missing data procedure. We have omitted, at random, 15% of the values of the 8-year maths score. Three analyses have been carried out. The first simply omits all the level 1 units with a missing value. The second carries out only the first stage of the analysis to provide a completed dataset and then proceeds in the usual way. The third analysis carries out the full missing data procedure.

The first stage consists of estimating the level 2 and level 3 covariance matrices for the response and three explanatory variables (excluding the intercept) and estimating the residuals.

We see that in the analysis which retains only the complete cases the standard errors tend to be raised. The analysis which uses the completed dataset without adjusting for the uncertainty of the predicted values tends to underestimate the level 1 variance and also changes the fixed parameter estimates. The corrected analysis using the full missing data procedure tends to give standard errors which are somewhat smaller than the analysis which simply omits level 1 units with missing data.

14.6   MCMC estimation for missing data

We can use MCMC estimation in two ways to handle missing data. In the first we use the same device as in section 14.2. Considering the set of variables as having a multivariate Normal distribution we fit the model using MCMC and then sample sets of residuals corresponding to the missing values. A convenient way of doing this is to run the chain, after convergence for, say, 1000 iterations, take the residuals for the final iteration, then run for another 1000 and take the final set etc. This will normally ensure that the sets of residuals are independent. With these sets of residuals we can use the same procedure of multiple imputation given by (14.5).

The second, more direct, approach using MCMC is to treat the missing data as further parameters to be estimated. This then involves an additional step in the algorithm whereby the missing values are sampled from their distribution given the current values of the other parameters. In the multivariate Normal case, if we place uniform priors on the missing data then the conditional distributions are multivariate Normal. This can be extended to handle categorical data and a discussion is given by Schafer (1997).

14.7   Informatively missing data

So far we have assumed that we can treat the missing values as randomly missing, possibly conditionally given other included variables. We now look at ways of handling the case when this assumption is unrealistic. Such situations can arise in a number of ways. We considered some of these in Chapter 5 for individuals dropping out of repeated measurement studies. More generally, individuals may fail to respond to surveys or to some questions in a survey. We shall not consider the latter case separately, assuming either that missingness can be considered random given variables that are known, as in the examples discussed in previous sections, or that these are individuals with no information who are treated as such. We are here concerned with missing data whose unknown values are associated with one or more random variables in the model.

Heckman (1979) described a framework for handling such data in single-level models. We write the following 2-level version of this model. Suppose the target model of interest is

$$y_{ij} = (X_1\beta_1)_{ij} + u_{1j} + e_{1ij} \tag{14.8}$$

and suppose that we also have a model for the probability of non-response

$$\pi_{ij} = g[(X_2\beta_2)_{ij} + u_{2j}] \tag{14.9}$$

where g is the probit function or possibly some other suitable link function such as the logit. The binary response is whether or not there is a response and it is assumed that the set of covariates used in (14.7), X_2, is available for all individuals, whether the response is observed or not. We assume, as in section 4.8, that (14.9) is derived from an underlying Normal model

$$z_{ij} = (X_2\beta_2)_{ij} + u_{2j} + e_{2ij}$$

and that a positive value, in other words an observed response occurs when $z_{ij} > 0$. In general, in order to satisfy identifiability requirements, at least one of the covariates in

(14.9) should not be included in (14.8) and all the covariates should be independent of the random effects, but see Little (1985) for a further discussion. Heckman suggested a two-step estimation procedure.

Model (14.9) is estimated first. We can then write down a Normal linear model for the observed responses, that is for $E(y_{ij}|z_{ij} > 0)$, which involves the parameters of interest in (14.8). This can obtained from the covariance structure of the joint model with level 2 and level 1 covariance matrices given by

$$\Omega_u = \begin{pmatrix} \sigma_{u1}^2 \\ \sigma_{u12} & \sigma_{u2}^2 \end{pmatrix}, \quad \Omega_e = \begin{pmatrix} \sigma_{e1}^2 \\ \sigma_{e12} & \sigma_{e2}^2 \end{pmatrix} \tag{14.10}$$

In the single-level case this reduces to a simple form.

A two-step procedure is not generally fully efficient and, in fact, the joint model given by (14.8)–(14.10) is simply a bivariate 2-level model which can be fitted efficiently as described in section 4.8, for example using MCMC estimation. Extensions to handle random coefficients and higher levels and more complex structures such as cross-classifications follow as with other models. The incorporation of higher level random effects will often be important in surveys where there are a limited number of suitable individual-level covariates for the response probability model, but informative cluster-level or interviewer-level variables may be available. The model can also be extended to multivariate response models by adding more response equations such as (14.8). The dropout models for repeated measures data discussed in Chapter 5 can be viewed as special cases of the present general model and a discussion is given by Crouchley and Ganjali (2002).

In the analysis of surveys, where suitable covariates for the response probability component model are unavailable, we may have unit weights that incorporate, at least partly, non-response adjustments. In this case we can carry out a weighted analysis. Since the weights, in general, are not independent of the random effects, this requires some care. Pfeffermann et al. (1997) discuss this in detail and suggest a series of likelihood-based approaches. Pfeffermann (2001) suggests an MCMC-based approach.

The simultaneous equations model of this section is one example where there is a target model of interest and auxiliary equations that are concerned with violations of assumptions. Such models are often referred to as *multiprocess* models. In the case of the missing data model the response is only observed if it takes certain values and we specify a second process for the missingness mechanism. Another example is that of 'endogeneity' where an unobserved or latent variable is associated with an explanatory variable and also with a random effect in the model and a second process model for this explanatory variable is specified. Steele (2001) gives an example using event history data. Another approach to the case where an explanatory variable is correlated with one or more random effects is given by Rice et al. (1998), who propose a procedure whereby residuals from the model are incorporated iteratively into the IGLS estimation, and where no explicit second process is specified.

15

Software for Multilevel Modelling, Resources and Further Developments

15.1 Software packages and resources

There are now many software packages that will carry out multilevel modelling. Most of the major statistical packages have features for the basic models, and can fit more complex or specialized models using a macro language. We shall not give a detailed review here, but merely list the main packages together with web addresses for further information. Recently published reviews of some of the packages are those of De Leeuw and Kreft (2001), Zhou et al. (1999) and Fein and Lissitz (2000). The Centre for Multilevel Modelling (www.multilevel.ioe.ac.uk) maintains a series of reviews.

Table 15.1 lists the packages, together with the internet address and a brief note.

The methodological literature on multilevel modelling is growing rapidly as is the literature on applications. The Centre for Multilevel Modelling endeavours to maintain a selection of the methodological literature and links to other resources such as websites and training materials. The Centre also produces a twice-yearly electronic newsletter with free subscription: this contains articles about current developments, reviews etc. A collection of datasets together with training materials and a version of the MLwiN package that will work with these datasets, is freely available at http://www.tramss.data-archive.ac.uk. Another useful resource for multilevel modelling is http://www.ats.ucla.edu/stat/mlm/default.htm.

There is a very active email discussion group that can be accessed and joined at http://www.jiscmail.ac.uk/lists/multilevel.html. The group serves as a means of exchanging information and suggestions about data analysis.

15.2 Further developments

While the models and estimation procedures described in preceding chapters are extensive there remain many important areas for future development. Some important ones are, briefly, as follows.

Table 15.1 Available software packages

Name	Web site	Note
aML	http://www.applied-ml.com	Concentrates on event history and multiprocess models. Maximum likelihood estimation
ASREML	http://www.vsn-intl.com/asreml/	Same features as GENSTAT
BAYESX	http://www.stat.uni-meunchen.de/~lang/bayesx/bayesx.html	General-purpose MCMC estimation. Continuous and discrete responses with nested and cross-classified structures. Concentrates on semiparametric regression
BMDP	http://www.statsol.ie/bmdp/bmdp.htm	Variance components model and serial correlations for nested structures. Maximum likelihood and GEE
EGRET	http://www.cytel.com/products/egret	Discrete responses for nested structures up to two levels. Maximum likelihood estimation
GENSTAT	http://www.nag.co.uk/stats/tt_soft.asp	Continuous and discrete responses. Nested and cross-classified structures. Maximum likelihood estimation
HLM	http://www.ssicentral.com/hlm	Continuous and discrete responses up to three levels. Serial correlation structures; measurement errors. Maximum and quasilikelihood estimation
LIMDEP	www.limdep.com	General-purpose econometric software. Continuous and discrete responses. Nested structures. Maximum likelihood estimation
LISREL	http://www.ssicentral.com/lisrel.htm	Multilevel structural equations. Nested data structures. Maximum likelihood estimation
MIXOR, MIXREG	http://tigger.uic.edu/~hedeker/mix.html	A suite of programs for continuous and discrete response multicategory models up to three levels. Maximum likelihood estimation
MLWiN	www.multilevel.ioe.ac.uk	General-purpose package. Continuous and discrete responses for nested, cross-classified and multiple membership structures for any number of levels. Serial correlation structures; event history models; factor analysis; measurement errors. Maximum and quasilikelihood estimation; MCMC estimation
MPLUS	http://www.statmodel.com/mplus	Continuous and discrete responses. Nested structures. Multilevel structural equations. Maximum likelihood estimation
OSWALD	http://www.maths.lancs.ac.uk/Software/Oswald/	Works with S PLUS for analysis of serial correlation and event history data. Maximum likelihood and GEE estimation

(Continued)

Table 15.1   (*Continued*)

Name	Web site	Note
SAS (version 8)	http://www.sas.com/products/ sassystem/release82/	General-purpose package. Continuous and discrete responses for nested and cross-classified structures up to two levels. Serial correlation structures. Maximum and quasilikelihood estimation
S-PLUS 2000	http://www.insightful.com	General-purpose software. Continuous and discrete responses for nested structures
SPSS (version 10)	http://www.spss.com/spss10	General-purpose package. Handles basic continuous and discrete response models for nested structures up to two levels. Maximum likelihood estimation
STATA	http://www.stata.com	General-purpose package. Continuous and discrete responses. Nested and cross-classified structures up to two levels. Structural equation models (GLLAMM program), serial correlations. Maximum likelihood estimation
SYSTAT	http://www.spssscience.com/systat	General-purpose package. Continuous responses. Nested structures up to two levels. Serial correlations. Maximum likelihood estimation
WINBUGS	http://www.mrc-bsu.cam.ac.uk	General-purpose package. Uses MCMC to fit a very wide range of models via a statistical control language. Continuous and discrete responses, measurement errors, factor analysis, serial correlations and more

Further work on missing data, especially where it is informatively missing, is needed, and sensitivity analysis to the various assumptions that have to be made would be useful. There is a considerable amount of work to be done on measurement error models, including errors at different levels of a data hierarchy and satisfactory ways of estimating measurement error distributions. Further work on the specification and fitting of structural equation models would be useful. More work on optimal design is needed as is work on diagnostics; in both cases a useful start has been made.

On the computational side, with an increasing use of very large datasets, ways of improving the efficiency of existing methods, especially for MCMC modelling, would be very useful, although the increasing power and memory capacity of computers will be of some help here.

While many of the procedures described in these chapters provide powerful tools for the exploration of complex data structures, in many areas there are few datasets that are collected in such a way that allows these models to be applied. Thus, information that identifies units may be absent, or research questions may have been phrased in ways that assume only single-level analysis is to be carried out. One of the most pressing needs, therefore, is for researchers to become familiar with multilevel modelling techniques and their possibilities. Although multilevel modelling is becoming more widely used and understood, there is still a need for good introductory materials and training generally and the introduction of courses into undergraduate and postgraduate teaching is welcome. Given the ubiquity of multilevel data, in time it should become a standard technique for data analysis in the same way that ordinary regression has been during the latter part of the twentieth century.

References

Abdous, B. and Berlinet, A. (1998). Pointwise improvement of multivariate kernel density estimates. *Journal of Multivariate Analysis*, **65**, 109–28.

Aitkin, M. and Longford, N. (1986). Statistical modelling in school effectiveness studies (with discussion). *Journal of the Royal Statistical Society*, A **149**, 1–43.

Aitkin, M., Anderson, D. and Hinde, J. (1981). Statistical modelling of data on teaching styles (with discussion). *Journal of the Royal Statistical Society*, A **144**, 148–61.

Aitkin, M., Anderson, D., Francis, B. and Hinde, J. (1989). *Statistical Modelling in GLIM*. Oxford, Clarendon Press.

Ansari, A. and Jedidi, K. (2002). Heterogeneous factor analysis models: a Bayesian approach. *Psychometrika*, **67**, 49–78.

Barbosa, M.F. and Goldstein, H. (2000). Discrete response multilevel models for repeated measures; an application to voting intentions data. *Quality and Quantity*, **34**, 323–30.

Beale, E.M.L. and Little, R.J.A. (1975). Missing values in multivariate analysis. *Journal of the Royal Statistical Society*, B **37**, 129–45.

Bennett, N. (1976). *Teaching Styles and Pupil Progress*. London, Open Books.

Blatchford, P., Goldstein, H., Martin, C. and Browne, W. (2002). A study of class size effects in English school reception year classes. *British Educational Research Journal*, **28**, 169–85.

Bock, R.D. (1992). *Structural and Nonstructural Analysis of Multiphasic Growth*. Chicago, University of Chicago (unpublished).

Box, G.E.P. and Cox, D.R. (1964). An analysis of transformations (with discussion). *Journal of the Royal Statistical Society*, B **26**, 211–52.

Breslow, N.E. and Clayton, D.G. (1993). Approximate inference in generalised linear mixed models. *Journal of the American Statistical Association*, **88**, 9–25.

Browne, W. (1998). *Applying MCMC Methods to Multilevel Models. Statistics*. Bath, University of Bath.

Browne, W. and Draper, D. (2000). Implementation and performance issues in the Bayesian and likelihood fitting of multilevel models. *Computational Statistics*, **15**, 391–420.

Browne, W., Draper, D., Goldstein, H. and Rasbash, J. (2001a). Bayesian and likelihood methods for fitting multilevel models with complex level 1 variation. Submitted.

Browne, W., Goldstein, H. and Rasbash, J. (2001b). Multiple membership multiple classification (MMMC) models. *Statistical Modelling*, **1**, 103–24.

Browne, W., Goldstein, H., Woodhouse, G. and Yang, M. (2001c). An MCMC algorithm for adjusting for errors in variables in random slopes multilevel models. *Multilevel Modelling Newsletter*, **13**(1), 4–9.

Browne, W.J. (2002). *MCMC Estimation in MLwiN*. London, Institute of Education.

Bryk, A.S. and Raudenbush, S.W. (2002). *Hierarchical Linear Models.* Newbury Park, California, Sage.

Burdick, R.K. and Graybill, F.A. (1988). The present status of confidence interval estimation on variance components in balanced and unbalanced random models. *Communications in Statistics: theory and methods*, **17**, 1165–95.

Burstein, L., Fischer, K.H. and Miller, M.D. (1980). The multilevel effects of background on science achievement: a cross national comparison. *Sociology of Education*, **53**, 215–25.

Bynner, J., Elias, P., McKnight, A., Pan, H., et al. (2002). *Young People in Transition: Changing Pathways to Employment and Independence.* York, Joseph Rowntree Foundation.

Carpenter, J., Goldstein, H. and Rasbash, J. (1999). A non-parametric bootstrap for multilevel models. *Multilevel Modelling Newsletter*, **11**, 2–5.

Carpenter, J., Goldstein, H. and Rasbash, J. (2002). A novel bootstrap procedure for assessing the relationship between class size and achievement. Submitted.

Clayton, D. and Kaldor, J. (1987). Empirical Bayes estimates of age-standardised relative risks for use in disease mapping. *Biometrics*, **43**, 671–81.

Clayton, D. and Rasbash, J. (1999). Estimation in large crossed random effect models by data augmentation. *Journal of the Royal Statistical Society*, A **162**, 425–36.

Clayton, D.G. (1988). The analysis of event history data: a review of progress and outstanding problems. *Statistics in Medicine*, **7**, 819–41.

Clayton, D.G. (1991). A Monte Carlo method for Bayesian inference in frailty models. *Biometrics*, **47**, 467–85.

Clayton, D.G. (1992). *Bayesian Analysis of Frailty Models.* Cambridge, MRC Biostatistics Unit (unpublished).

Cochran, W.G. (1983). *Planning and Analysis of Observational Studies.* New York, Wiley.

Cohen, M.P. (1998). Determining sample sizes for surveys with data analysed by hierarchical linear models. *Journal of Official Statistics*, **14**, 267–75.

Cole, T.J. and Green, P.J. (1992). Smoothing reference centile curves: the LMS method and penalized likelihood. *Statistics in Medicine*, **11**, 1305–19.

Cox, D.R. (1972). Regression models and life tables (with discussion). *Journal of the Royal Statistical Society*, B **34**, 187–220.

Cox, D.R. and Oakes, D. (1984). *Analysis of Survival Data.* London, Chapman and Hall.

Creswell, M. (1991). A multilevel Bivariate Model. In: Prosser, R., Rasbash, J. and Goldstein, H., eds. *Data Analysis with ML3.* London, Institute of Education.

Cronbach, L.J. and Webb, N. (1975). Between class and within class effects in a repeated aptitude × treatment interaction: reanalysis of a study by G.L. Anderson. *Journal of Educational Psychology*, **67**, 717–24.

Crouchley, R. and Ganjali, M. (2002). The common structure of several models for non-ignorable dropout. *Statistical Modelling*, **2**, 39–62.

Davison, A.C. and Hinkley, D.V. (1997). *Bootstrap Methods and their Application.* Cambridge, Cambridge University Press.

De Leeuw, J. and Kreft, I. (2001). Software for multilevel analysis. In: Leyland, A. and Goldstein, H., eds. *Multilevel Modelling of Health Statistics.* Chichester, Wiley.

Demirjian, A., La Palme, L. and Thibault, H.W. (1982). La croissance staturo-pondérale des enfants Canadien-Français de la naissance à 36 mois. *Union Médicale du Canada*, **112**, 153–63.

Derbyshire, M.E. (1987). Statistical rationale for grant-related expenditure assessment (GREA) concerning personal social services. *Journal of the Royal Statistical Society*, A **150**, 309–33.

Diggle, P. and Kenward, M.G. (1994). Informative dropout in longitudinal data analysis. *Applied Statistics*, **43**, 49–93.

Diggle, P.J. (1988). An approach to the analysis of repeated measurements. *Biometrics*, **44**, 959–71.

Draper, D. (2002). *Bayesian Hierarchical Modelling.* New York, Springer-Verlag.

Ecob, R. and Goldstein, H. (1983). Instrumental variable methods for the estimation of test score reliability. *Journal of Educational Statistics,* **8**, 223–41.

Efron, B. (1988). Logistic regression, survival analysis, and the Kaplan–Meier curve. *Journal of the American Statistical Association,* **83**, 414–25.

Efron, B. and Gong, G. (1983). A leisurely look at the Bootstrap, the Jacknife and Cross-validation. *The American Statistician,* **37**, 36–48.

Egger, P.J. (1992). Event history analysis: discrete-time models including unobserved heterogeneity, with applications to birth history data. University of Southampton, PhD thesis.

Everitt, B. (1984). *Introduction to Latent Variable Models.* London, Chapman and Hall:

Fein, M. and Lissitz, R.W. (2000). *Comparison of HLM and MLwiN Multilevel Analysis Software Packages: a Monte Carlo Investigation into the Equality of the Estimates.* University of Maryland.

Fuller, W.A. (1987). *Measurement Error Models.* New York, Wiley.

Garrett, M., Fitzmaurice, M. and Laird, N. (1993). A likelihood based method for analysing longitudinal binary responses. *Biometrika,* **80**, 141–51.

Gelfand, A.E., Sahu, S.K. and Carlin, B.P. (1995). Efficient parameterisations for Normal linear mixed models. *Biometrika,* **82**, 479–88.

Gelman, A., Roberts, G.O. and Gilks, W.R. (1996). Efficient Metropolis jumping rules. *Bayesian Statistics,* **5**, 599–607.

Gilks, W.R. and Wild, P. (1992). Adaptive rejection sampling for Gibbs sampling. *Applied Statistics,* **41**, 337–48.

Gilks, W.R., Richardson, S. and Spiegelhalter, D.J. (1996). *Markov Chain Monte Carlo in Practice.* London, Chapman and Hall:

Goldstein, H. (1976). Smoking in pregnancy: some notes on the statistical controversy. *British Journal of Preventive and Social Medicine,* **31**, 13–17.

Goldstein, H. (1979). *The Design and Analysis of Longitudinal Studies.* London, Academic Press.

Goldstein, H. (1986). Multilevel mixed linear model analysis using iterative generalised least squares. *Biometrika,* **73**, 43–56.

Goldstein, H. (1987a). Multilevel covariance component models. *Biometrika,* **74**, 430–31.

Goldstein, H. (1987b). *Multilevel Models in Educational and Social Research.* London, Griffin.

Goldstein, H. (1987c). The choice of constraints in correspondence analysis. *Psychometrika,* **52**, 207–15.

Goldstein, H. (1989a). Restricted unbiased iterative generalised least squares estimation. *Biometrika,* **76**, 622–23.

Goldstein, H. (1989b). Efficient prediction models for adult height. In: Tanner, J.M., ed. *Auxology 88; Perspectives in the science of growth and development.* London, Smith Gordon, pp. 41–48.

Goldstein, H. (1991). Nonlinear multilevel models with an application to discrete response data. *Biometrika,* **78**, 45–51.

Goldstein, H. (1992). Statistical information and the measurement of education outcomes (editorial). *Journal of the Royal Statistical Society,* A **155**, 313–15.

Goldstein, H. (1995). *Multilevel Statistical Models,* 2nd edition. London, Edward Arnold: New York, Wiley.

Goldstein, H. (1997). Methods in school effectiveness research. *School Effectiveness and School Improvement,* **8**, 369–95.

Goldstein, H. (1998). *Models for Reality.* London, Institute of Education.

Goldstein, H. and Blatchford, P. (1998). Class size and educational achievement: a review of methodology with particular reference to study design. *British Educational Research Journal* **24**, 255–68.

Goldstein, H. and Browne, W. (2002a). Binary response factor modelling of achievement data. In: Olivares, A. and McArdle, J., eds. *Psychometrics: a Festschrift to Roderick P. McDonald*. Hillsdale, New Jersey, Lawrence Erlbaum.

Goldstein, H. and Browne, W. (2002b). Multilevel factor analysis modelling using Markov Chain Monte Carlo estimation. In: Marcoulides, G. and Moustaki, I., eds. *Latent Variable and Latent Structure Models*. London, Lawrence Erlbaum, pp. 225–244.

Goldstein, H., and Healy, M.J.R. (1995). The graphical presentation of a collection of means. *Journal of the Royal Statistical Society*, A **158**, 175–7.

Goldstein, H. and McDonald, R.P. (1987). A general model for the analysis of multilevel data. *Psychometrika*, **53**, 455–67.

Goldstein, H. and Rasbash, J. (1992). Efficient computational procedures for the estimation of parameters in multilevel models based on iterative generalised least squares. *Computational Statistics and Data Analysis*, **13**, 63–71.

Goldstein, H. and Rasbash, J. (1996). Improved approximations for multilevel models with binary responses. *Journal of the Royal Statistical Society*, A **159**, 505–13.

Goldstein, H. and Spiegelhalter, D.J. (1996). League tables and their limitations: statistical issues in comparisons of institutional performance. *Journal of the Royal Statistical Society*, A **159**, 385–443.

Goldstein, H. and Wood, R. (1989). Five decades of item response modelling. *British Journal of Mathematical and Statistical Psychology*, **42**, 139–67.

Goldstein, H., Rasbash, J., Yang, M., Woodhouse, G., et al. (1993). A multilevel analysis of school examination results. *Oxford Review of Education*, **19**, 425–33.

Goldstein, H., Healy, M.J.R. and Rasbash, J. (1994). Multilevel time series models with applications to repeated measures data. *Statistics in Medicine*, **13**, 1643–55.

Goldstein, H., Rasbash, J., Browne, W., Woodhouse, G., et al. (2000a). Multilevel models in the study of dynamic household structures. *European Journal of Population*, **16**, 373–87.

Goldstein, H., Yang, M., Omar, R., Turner, R., et al. (2000b). Meta analysis using multilevel models with an application to the study of class size effects. *Journal of the Royal Statistical Society*, C **49**, 399–412.

Goldstein, H., Browne, W. and Rasbash, J. (2002a). Partitioning variation in multilevel models. *Understanding Statistics*, 2, in press.

Goldstein, H., Pan, H. and Bynner, J. (2002b). A note on methodology for analysing longitudinal event histories using repeated partnership data from the National Child Development Study. Submitted.

Greenacre, M.J. (1984). *Theory and Applications of Correspondence Analysis*. New York, Academic Press.

Grizzle, J.C. and Allen, D.M. (1969). An analysis of growth and dose response curves. *Biometrics*, **25**, 357–61.

Gumpertz, M.L. and Pantula, S.G. (1992). Nonlinear regression with variance components. *Journal of the American Statistical Association*, **87**, 201–9.

Harrison, G.A. and Brush, G. (1990). On correlations between adjacent velocities and accelerations in longitudinal growth data. *Annals of Human Biology*, **17**, 55–7.

Hartzel, J., Agresti, A. and Caffo, B. (2001). Multinomial logit random effects models. *Statistical Modelling*, **1**, 81–102.

Heagerty, P.J. and Zeger, S.L. (2000). Marginalized multilevel models and likelihood inference (with discussion). *Statistical Science*, **15**, 1–26.

Heath, A., Yang, M. and Goldstein, H. (1996). Multilevel analysis of the changing relationship between class and party in Britain 1964–1992. *Quality and Quantity*, **30**, 389–404.

Heck, R.H. and Thomas, S.L. (2000). *An Introduction to Multilevel Modelling Techniques*. Mahwah, New Jersey, Lawrence Erlbaum.

Heckman, J.J. (1979). Sample selection bias as a specification error. *Econometrica*, **47**, 153–61.

Hedeker, D. and Gibbons, R.D. (1994). A random effects ordinal regression model for multilevel analysis. *Biometrics*, **50**, 933–44.

Hedges, L.V. and Olkin, I.O. (1985). *Statistical Methods for Meta Analysis*. Orlando, Florida, Academic Press.

Hill, P.W. and Goldstein, H. (1998). Multilevel modelling of educational data with cross classification and missing identification of units. *Journal of Educational and Behavioural Statistics*, **23**, 117 28.

Hodges, J. (1998). Some algebra and geometry for hierarchical models, applied to diagnostics. *Journal of the Royal Statistical Society*, B **60**, 497–536.

Holland, P.W. (1986). Statistics and causal inference. *Journal of the American Statistical Association*, **81**, 945–71.

Hox, J. (2002). *Multilevel Analysis, Techniques and Applications*. Mahwah, New Jersey, Lawrence Erlbaum Associates.

Huq, N.M. and Cleland, J. (1990). *Bangladesh Fertility Survey 1989*. Dhaka, National Institute of Population Research and Training.

Jenss, R.M. and Bayley, N. (1937). A mathematical method for studying the growth of a child. *Human Biology*, **9**, 556–63.

Joreskog, K.G. and Sorbom, D. (1979). *Advances in Factor Analysis and Structural Equation Models*. Cambridge, Massachusetts, Abt books.

Kass, R.E. and Steffey, D. (1989). Approximate Bayesian inference in conditionally independent hierarchical models (parametric empirical Bayes models). *Journal of the American Statistical Association*, **84**, 717–26.

Kenward, M. (1998). Selection models for repeated measurements with non random dropout: an illustration of sensitivity. *Statistics in Medicine*, **17**, 2723–32.

Kish, L. (1965). *Survey Sampling*. New York, Wiley.

Kish, L. and Frankel, M.R. (1974). Inference from complex samples. *Journal of the Royal Statistical Society*, B **36**, 1–37.

Kreft, I. and De Leeuw, J. (1998). *Introducing Multilevel Modelling*. London, Sage.

Kuk, A.Y.C. (1995). Asymptotically unbiased estimation in generalised linear models with random effects. *Journal of the Royal Statistical Society*, B **57**, 395–407.

Laird, N.M. (1988). Missing data in longitudinal studies. *Statistics in Medicine*, **7**, 305–15.

Laird, N.M. and Louis, T.A. (1987). Empirical Bayes confidence intervals based on bootstrap samples. *Journal of the American Statistical Association*, **82**, 739–57.

Laird, N.M. and Louis, T.A. (1989). Empirical Bayes confidence intervals for a series of related experiments. *Biometrics*, **45**, 481–95.

Langford, I. and Lewis, T. (1998). Outliers in multilevel data. *Journal of the Royal Statistical Society*, A **161**, 121–60.

Langford, I., Leyland, A.H., Rasbash, J. and Goldstein, H. (1999). Multilevel modelling of the geographical distributions of diseases. *Journal of the Royal Statistical Society*, C **48**, 253–68.

Larsen, U. and Vaupel, J.W. (1993). Hutterite fecundability by age and parity: strategies for frailty modelling of event histories. *Demography*, **30**, 81–101.

Lawley, D.N. and Maxwell, A.E. (1971). *Factor Analysis as a Statistical Method*, 2nd edition. London, Butterworth.

Lee, Y. and Nelder, J.A. (1996). Hierarchical generalised linear models. *Journal of the Royal Statistical Society*, B **58**, 619–78.

Lee, Y. and Nelder, J.A. (2001). Hierarchical generalised linear models: a synthesis of generalised linear models, random effect models and structured dispersions. *Biometrika*, **88**, 987–1006.

Lehtonen, R. and Veijanen, A. (1999). *Multilevel Model Assisted Generalised Regression Estimators for Domain Estimation*. International Association of Survey Statisticians Workshop, Riga.

Lesaffre, E. and Spiessens, B. (2001). On the effect of the number of quadrature points in a logistic random effects model: an example. *Journal of the Royal Statistical Society*, C **50**, 325–36.

Lewis, T. and Langford, I. (2001). Outliers, robustness and the detection of discrepant data. In: Leyland, A. and Goldstein, H., eds. *Multilevel Modelling of Health Statistics*. Chichester, Wiley, pp. 75–92.

Liang, K.-Y., Zeger, S.L. and Qaqish, B. (1992). Multivariate regression analyses for categorical data. *Journal of the Royal Statistical Society*, B **54**, 3–40.

Lindsey, J.K. (1999). Relationships among sample size, model selection and likelihood regions, and scientifically important differences. *Journal of the Royal Statistical Society*, D **48**, 401–12.

Lindsey, J.K. and Lambert, P. (1998). On the appropriateness of marginal models for repeated measurements in clinical trials. *Statistics in Medicine*, **17**, 447–69.

Lindstrom, M.J. and Bates, D.M. (1990). Nonlinear mixed effects models for repeated measures data. *Biometrics*, **46**, 673–87.

Little, R. (1985). A note about models for selectivity bias. *Econometrica*, **53**, 1469–74.

Little, R.J.A. (1992). Regression with missing X's: a review. *Journal of the American Statistical Association*, **87**, 1227–37.

Little, T.D., Schnabel, K.U. and Baumert, J., eds (2000). *Modelling Longitudinal and Multilevel Data*. New Jersey, Lawrence Erlbaum.

Liu, Q. and Pierce, D.A. (1994). A note on Gauss–Hermite quadrature. *Biometrika*, **81**, 13–22.

Longford, N. (1999). Multivariate shrinkage estimation of small area means and proportions. *Journal of the Royal Statistical Society*, A **162**, 227–45.

Longford, N.T. (1993). *Random Coefficient Models*. Oxford, Clarendon Press.

Longford, N.T. (1987). A fast scoring algorithm for maximum likelihood estimation in unbalanced mixed models with nested random effects. *Biometrika*, **74**, 817–27.

Longford, N.T. and Muthen, B.O. (1992). Factor analysis for clustered populations. *Psychometrika*, **57**, 581–97.

McCullagh, P. and Nelder, J. (1989). *Generalised Linear Models*, 2nd edition. London, Chapman and Hall.

McCulloch, C.E. (1997). Maximum likelihood algorithms for generalised linear mixed models. *Journal of the American Statistical Association*, **92**, 162–70.

McCulloch, C.E. and Searle, S.R. (2001). *Generalised, Linear and Mixed Models*. New York, Wiley.

McDonald, R.P. (1985). *Factor Analysis and Related Methods*. Hillsdale, New York, Lawrence Erlbaum.

McDonald, R.P. (1993). A general model for two level data with responses missing at random. *Psychometrika*, **58**, 575–85.

McDonald, R.P. and Goldstein, H. (1988). Balanced versus unbalanced designs for linear structural relations in two level data. *British Journal of Mathematical and Statistical Psychology*, **42**, 215–32.

McGrath, K. and Waterton, J. (1986). *British Social Attitudes, 1983–1986 panel survey*. London, Social and Community Planning Research.

Meng, X. and Dyk, D. (1998). Fast EM implementations for mixed effects models. *Journal of the Royal Statistical Society*, B **60**, 559–78.

Miller, R.G. (1974). The Jacknife—a review. *Biometrika*, **61**, 1–15.

Moerbeek, M. and Wong, W.K. (2002). Multiple objective optimal designs for the hierarchical linear model. Submitted.

Moerbeek, M., Van Breukelen, G.J.P. and Berger, M.P.F. (2000). Design issues for experiments in multilevel populations. *Journal of Educational and Behavioural Statistics*, **25**, 271–84.

Moerbeek, M., Van Breukelen, J.P. and Berger, M.P. (2001). Optimal experimental design for multilevel logistic models. *Journal of the Royal Statistical Society*, D **50**, 17–30.

Mok, M. (1995). Sample size requirements for 2-level designs in educational research. *Multilevel Modelling Newsletter*, **7**(2), 11–15.

Mortimore, P., Sammons, P., Stoll, L., Lewis, D. and Ecob, R. (1988). *School Matters*. Wells, Open Books.

Muthen, B.O. (1989). Latent variable modelling in heterogeneous populations. *Psychometrika*, **54**, 557–85.

Nuttall, D.L., Goldstein, H., Prosser, R. and Rasbash, J. (1989). Differential school effectiveness *International Journal of Educational Research*, **13**, 769–76.

Pan, H. and Goldstein, H. (1997). Multilevel models for longitudinal growth norms. *Statistics in Medicine*, **16**, 2665–78.

Pan, H. and Goldstein, H. (1998). Multilevel repeated measures growth modelling using extended spline functions. *Statistics in Medicine*, **17**, 2755–70.

Paterson, L. (1991). Socio-economic status and educational attainment: a multidimensional and multilevel study. *Evaluation and Research in Education*, **5**, 97–121.

Peto, R. (1972). Contribution to discussion of paper by D.R. Cox. *Journal of the Royal Statistical Society*, B **34**, 205–7.

Pfeffermann, D. (2001). *Multilevel Modelling under Informative Probability Sampling*. 53rd Session of International Statistical Institute, Seoul.

Pfeffermann, D., Skinner, C.J., Holmes, D., et al. (1997). Weighting for unequal selection probabilities in multilevel models. *Journal of the Royal Statistical Society*, B **60**, 23–40.

Plewis, I. (1996). Statistical methods for understanding cognitive growth: a review, a synthesis and an application. *British Journal of Mathematical and Statistical Psychology*, **49**, 25–42.

Plewis, I. (1985). *Analysing Change*. Chichester, Wiley.

Plewis, I. (1993). Reading progress. In: Woodhouse, G., ed. *A Guide to ML3 for New Users*. London, Multilevel Models Project.

Pourahmadi, M. (1999). Joint mean-covariance models with applications to longitudinal data: unconstrained parameterisation. *Biometrika*, **86**, 677–90.

Pourahmadi, M. (2000). Maximum likelihood estimation of generalised linear models for multivariate Normal covariance matrix. *Biometrika*, **87**, 425–36.

Putt, M. and Chinchilli, V.M. (1999). A mixed effects model for the analysis of repeated measures cross-over studies. *Statistics in Medicine*, **18**, 3037–58.

Rabe-Hesketh, S., Pickles, A. and Skrondal, A. (2001). GLLAMM: a general class of multilevel models and a STATA program. *Multilevel Modelling Newsletter*, **13**(1), 17–23.

Raftery, A.E. (1995). Bayesian model selection in social research. *Sociological Methodology*, **25**, 111–63.

Raftery, A.E. and Lewis, S.M. (1992). How many iterations in the Gibbs sampler? In: Bernardo, J.M., Berger, J.O., Dawid, A.P. and Smith, A.F.M., eds. *Bayesian Statistics 4*. Oxford, Oxford University Press, pp. 765–76.

Rasbash, J. and Goldstein, H. (1994). Efficient analysis of mixed hierarchical and cross classified random structures using a multilevel model. *Journal of Educational and Behavioural Statistics*, **19**, 337–50.

Rasbash, J., Browne, W., Goldstein, H., et al. (2000). *A User's Guide to MLwiN*, 2nd edition. London, Institute of Education.

Raudenbush, S.W. (1993). A crossed random effects model for unbalanced data with applications in cross-sectional and longitudinal research. *Journal of Educational Statistics*, **18**, 321–49.

Raudenbush, S.W. (1994). Equivalence of fisher scoring to iterative generalised least squares in the Normal case with application to hierarchical linear models. Unpublished.

Raudenbush, S.W. (1995). Maximum likelihood estimation for unbalanced multilevel covariance structure models via the EM algorithm. *British Journal of Mathematical and Statistical Psychology*, **48**, 359–70.

Rice, N., Jones, A. and Goldstein, H. (1998). *Multilevel Models where the Random Effects are Correlated with the Fixed Predictors: a conditioned iterative generalised least squares estimator (CIGLS).* York, University of York, Centre for Health Economics.

Robinson, W.S. (1950). Ecological correlations and the behaviour of individuals. *American Sociological Review*, **15**, 351–57.

Rodriguez, G. and Goldman, N. (2001). Improved estimation procedures for multilevel models with binary response: a case study. *Journal of the Royal Statistical Society*, A **164**, 339–56.

Rosier, M.J. (1987). The second international science study. *Comparative Education Review*, **31**, 106–28.

Rowe, K.J. and Hill, P.W. (1997). *Simultaneous Estimation of Multilevel Structural Equations to Model Students' Educational Progress.* Tenth International Congress for School Effectiveness and Improvement, Memphis, Tennessee.

Royall, R.M. (1986). Model robust confidence intervals using maximum likelihood estimators. *International Statistical Review*, **54**, 221–26.

Rubin, D.B. (1987). *Multiple Imputation for Nonresponse in Surveys.* New York, Wiley.

Rubin, D.B. and Thayer, D.T. (1982). EM algorithms for ML factor analysis. *Psychometrika*, **47**, 69–76.

Sarndal, C.-E., Swensson, B. and Wretman, J.H. (1992). *Model Assisted Survey Sampling.* New York, Springer.

Schafer, J.L. (1997). *Analysis of Incomplete Multivariate Data.* London, Chapman and Hall.

Searle, S.R., Casella, G. and McCulloch, C.E. (1992). *Variance Components.* New York, Wiley.

Self, S.G. and Liang, K. (1987). Asymptotic properties of maximum likelihood estimators and likelihood ratio tests under non-standard conditions. *Journal of the American Statistical Association*, **82**, 605–10.

Seltzer, M.H. (1993). Sensitivity analysis for fixed effects in the hierarchical model: a Gibbs sampling approach. *Journal of Educational Statistics*, **18**, 207–36.

Silverman, B. (1986). *Density Estimation for Statistics and Data Analysis.* London, Chapman and Hall.

Singer, J.D. and Willett, J.B. (2002). *Applied Longitudinal Data Analysis: modelling change and event occurrence.* New York, Oxford University Press.

Skinner, C.J., Holt, D. and Smith, T.M.F. (1989). *Analysis of Complex Surveys.* Chichester, Wiley.

Snijders, T.A.B. and Bosker, R.J. (1993). Standard errors and sample sizes for two-level research. *Journal of Educational Statistics*, **18**, 237–59.

Snijders, T. and Bosker, R. (1999). *Multilevel Analysis.* London, Sage.

Spiegelhalter, D.J., Best, N.G., Carlin, B.P. and Van der Linde, A. (2002). Bayesian measures of model complexity and fit. *Journal of the Royal Statistical Society*, B. In press.

Spiegelhalter, D.J., Thomas, A. and Best, N.G. (2000). *WinBUGS Version 1.3: user manual.* Cambridge, MRC Biostatistics Research Unit.

Steele, F. (2001). *A Multiprocess Multilevel Model to Allow for Selection Effects of Source of Contraceptive Supply in an Analysis of Contraceptive Discontinuation in Morocco.* 53rd Session of the International Statistical Institute, Seoul.

Steele, F., Goldstein, H. and Browne, W. (2003). A general multilevel multistate competing risks model for event history data. Submitted.

Tanner, M. and Wong, W.H. (1987). The calculation of posterior distributions by data augmentation. *Journal of the American Statistical Association*, **82**, 528–40.

Touloumi, G., Pocock, S.J., Babiker, A.G. and Darbyshire, J.H. (1999). Estimation and comparison of rates of change in longitudinal studies with informative dropouts. *Statistics in Medicine*, **18**, 1215–33.

Turner, R.M., Omar, R.Z., Yang, M., et al. (2000). A multilevel model framework for meta analysis of clinical trials with binary outcomes. *Statistics in Medicine*, **19**, 3417–32.

Vevea, J. (1994). A model for estimating effect size in the presence of publication bias. Paper presented to American Educational Research Association, annual meeting, New Orleans, April 1994.

Waclawiw, M.A. and Liang, K. (1994). Empirical Bayes estimation and inference for the random effects model with binary response. *Statistics in Medicine*, **13**, 541–51.

Wei, L.J., Lin, D.Y. and Weissfeld, L. (1989). Regression analysis of multivariate incomplete failure time data by modelling marginal distributions. *Journal of American Statistical Association*, **84**, 1065–73.

Wolfinger, R. (1993). Laplace's approximation for nonlinear mixed models. *Biometrika*, **80**, 791–5.

Woodhouse, G. (1998). *Adjustment for Measurement Error in Multilevel Analysis*. London, Institute of Education, University of London.

Woodhouse, G. and Goldstein, H. (1989). Educational Performance Indicators and LEA league tables. *Oxford Review of Education*, **14**, 301–19.

Woodhouse, G., Yang, M., Goldstein, H. and Rasbash, J. (1996). Adjusting for measurement error in multilevel analysis. *Journal of the Royal Statistical Society*, A **159**, 201–12.

Yang, M., Goldstein, H., Rath, T. and Hill, N. (1999). The use of assessment data for school improvement purposes. *Oxford Review of Education*, **25**, 469–83.

Yang, M., Goldstein, H. and Heath, A. (2000). Multilevel models for repeated binary outcomes: attitudes and vote over the electoral cycle. *Journal of the Royal Statistical Society*, A **163**, 49–62.

Yang, M., Goldstein, H., Browne, W. and Woodhouse, G. (2001). Multivariate multilevel analyses of examination results. *Journal of the Royal Statistical Society*, A **165**, 137–53.

Zeger, S.L., Liang, K-Y. and Albert, P.S. (1988). Models for longitudinal data: a generalised estimating equation approach. *Biometrics*, **44**, 1049–60.

Zhou, X., Perkins, A.J. and Hui, S.L. (1999). Comparisons of software packages for generalized linear multilevel models. *American Statistician*, **53**, 282–90.

Author Index

Subject Index